CW00726404

THE CHALLENGE OF THE THIRD REICH

THE CHALLENGE OF THE THIRD REICH

The Adam von Trott Memorial Lectures

Edited by

HEDLEY BULL

CLARENDON PRESS · OXFORD

1986

Oxford University Press, Walton Street, Oxford OX2 6DP
Oxford New York Toronto
Delhi Bombay Calcutta Madras Karachi
Kuala Lumpur Singapore Hong Kong Tokyo
Nairobi Dar es Salaam Cape Town
Melbourne Auckland
and associated companies in
Beirut Berlin Ibadan Nicosia

Oxford is a trade mark of Oxford University Press

Published in the United States
by Oxford University Press, New York

British Library Cataloguing in Publication Data
Bull, Hedley
The Challenge of the Third Reich : the Adam von Trott memorial lectures.
1. Germany—History—1933–1945
I. Title
943.086 DD256.5
ISBN 0-19-821962-8

Library of Congress Cataloging in Publication Data
The Challenge of the Third Reich.
Bibliography: p.
Includes index.
1. Germany—History—1933–1945—Addresses, essays, lectures. 2. World politics—1945— —Addresses, essays, lectures. I. Bull, Hedley. II. Title: Adam von Trott memorial lectures.
DD256.5.C545 1986 943.086 85-18763
ISBN 0-19-821962-8

Printed in Great Britain
at the University Press, Oxford
by David Stanford
Printer to the University

CONTENTS

LIST OF CONTRIBUTORS

DAVID ASTOR was Editor of *The Observer* from 1948 to 1975.

KARL-DIETRICH BRACHER is Professor at the Seminar für Politische Wissenschaft of Bonn University

MARTIN BROSZAT is Director of the Institut für Zeitgeschichte in Munich.

The late HEDLEY BULL, FBA, was Montague Burton Professor of International Relations at Oxford University, and Fellow of Balliol College.

KLEMENS VON KLEMPERER is L. Clark Seelye Professor of History at Smith College, Massachusetts.

PETER LUDLOW is Director of the Centre for European Policy Studies in Brussels.

TIMOTHY MASON is Emeritus Fellow of St Peter's College, Oxford.

HANS MOMMSEN is Professor of the Department of the Science of History at the Ruhr University, Bochum.

1

INTRODUCTION: THE CHALLENGE OF THE THIRD REICH

HEDLEY BULL

I

ADAM VON TROTT ZU SOLZ took part in the German resistance against Hitler and was executed at Plötzensee prison in Berlin on 26 August 1944 after the failure of the 20 July plot. He had been a Rhodes Scholar at Balliol College, where his name is inscribed outside the chapel in the list of old members who lost their lives in the Second World War.

The College decided to honour Adam von Trott's name with a series of lectures, following a suggestion from Professor Friedrich Wilhelm and an offer of financial help from his friend and contemporary at Balliol, David Astor. The lectures were held in Balliol Hall in Hilary Term 1983, coinciding with the fiftieth anniversary of Hitler's rise to power, and form the basis of the present volume.[1]

In calling these lectures *The Challenge of the Third Reich* we had in mind the challenge presented to the young Adam von Trott of how to respond to the regime that had taken power in his country, but also a series of other challenges: to the German people, of how to react to Nazi rule and to Nazi policies at home and abroad; to those in Germany, Trott among them, who chose to resist a ruthless police state, without significant popular support or encouragement from outside Germany and laying themselves open to the charge of treason, of how to express their resistance; and to Britain and the international community, of how to deal with a power committed to aggressive expansion, denial of the most basic human rights, and ultimately to genocide. We had in mind also the challenge presented to ourselves, half a century afterwards, to come to terms with the experience of the Third Reich, to consider again what meaning it has for the history of the twentieth century and what bearing it has on our own times.

[1] Karl-Dietrich Bracher was not able to deliver his lecture, owing to illness, but the text is included. Marion Gräfin Dönhoff gave a lecture on 'The Ethics of Resistance', but the text is not included.

II

Adam von Trott is by no means universally honoured as a martyr of anti-Nazi resistance either in Britain or, until recently, in Germany, but has been the subject of persistent controversy. While he worked secretly for the overthrow of Hitler his public role as an official in the German Foreign Office (and especially his frequent visits to neutral countries during the war) gave rise to the suspicion in Britain and the United States that, consciously or unconsciously, he was a Nazi agent. It is clear that he rejected both the means and the ends of Hitler's foreign policy and embraced a European and internationalist outlook, but along with his fellow conspirators and most Germans of his generation he accepted some of the revisionist aims of German foreign policy, and was regarded by some of his British acquaintances as a German nationalist. His own political leaning was that of a social democrat and anti-imperialist, and the resistance movement in which he played a part included a wide spectrum of points of view: its leadership and military arm, however, were conservative and aristocratic, and embodied values which, if they were anti-Nazi, were remote from those either of the Federal Republic of Germany or of the German Democratic Republic today. The claims of the anti-Nazi movement or movements in the Third Reich to represent an 'anderes Deutschland' with which the Allies could and should have been prepared to do business are no more widely accepted today in Britain and other Western countries than they were at the time.

But it is not possible to doubt that Adam von Trott was moved by a high sense of moral responsibility for the direction events had taken in his country, that he was opposed to the Nazi regime from the beginning, that he was committed to peace and reconciliation in Europe, or that he was a deeply courageous man. His own response to the challenge of the Third Reich—ineffectual as it was in deflecting the course of events, reached only after some hesitation, and controversial in the political premises on which it was based—was a noble one. In a personal tribute read during the lecture series Sir Isaiah Berlin concluded his remarks as follows: 'When researchers write to me and ask about Adam von Trott's life and political activities I can give them little concrete information: I can say no more than that he was he a brave and honourable man, a passionate patriot, incapable of anything ignoble or unworthy, and that he served what he regarded as being the deepest interests of his own countrymen and of decent people everywhere.'

III

The response of the German people as a whole to the challenge of the Third Reich cannot be regarded as a morally adequate one, for all that can be said to indicate the limits of their support for Nazi rule, to explain or excuse it, or to set it in the context of the moral failures and shortcomings of other peoples.

The German people during the period of the Third Reich cannot, of course, be identified with the acts of the Nazi regime. The fanatical followers of the Nazi ideology, as Martin Broszat makes clear below, were never more than a small minority. A substantial potential opposition was crushed at the outset in the brutal annihilation of communist, social democrat, and independent trades union dissidents described by Timothy Mason. While the mass of the German people responded enthusiastically to that part of Hitler's programme that called for national recovery, revision of the Versailles Treaty, and Greater Germany, schemes for a German empire in the east and its colonization by German settlers do not appear to have aroused much support. While Hitler's easy victories of 1940–1 were naturally intoxicating, the German people in the last years of peace were not driven by militarism or the glorification of war, and when war eventually came, they accepted it without enthusiasm. While anti-Semitism was an important element in German (and Austrian) political consciousness before the Nazi *Machtergreifung*, and racial hatred and theories of racial supremacy were always at the heart of Nazi thinking, the latter were not basic sources of the Nazi party's appeal. The majority of the German people had no direct part in the policies first of enforced deportation, later of mass extermination of European Jews. Terrible as the responsibility is that they bear for this most monstrous of Nazi crimes, for which Hitler himself may have tried to evade direct responsibility (Hans Mommsen's analysis below of the Holocaust takes account of the involvement in it not only of the SS but of the army, the railways, the banks, administration, and police; the moral indifference displayed by many Germans as Nazi anti-Jewish policy unfolded; their collective repression of the evidence that the Holocaust was taking place; the pursuit of bureaucratic efficiency in a moral vacuum; and the underlying weakness in the German public mind of the idea of man, which made all this possible), this is not the same as saying that these crimes expressed the will of the German people.

We have also to bear in mind the factors, some of them set out in this book, which help to account for German public support for Nazi rule or acquiescence in it. Hitler came to power by constitutional means, and went on to effect a 'legal revolution' which, as Karl-Dietrich Bracher argues, left the public confused, as well as disappointing the expectations of conservatives who had hoped to use the Nazi leader for their own purposes. Hitler's policies of deficit spending (Hitler, as Martin Broszat puts it, 'stumbled upon modern economics') brought about a remarkable economic recovery, with which his regime came henceforth to be identified. The regime stood for a new, more egalitarian and mobile social order, answering to the needs of an industrial society more effectively than the old order had done. At the same time it satisfied deep cravings for social unity and national self-assertion, frustrated in the years since 1918. Hitler's foreign policy successes in the late 1930s, like his military successes in the early 1940s, dramatic achievements which not only destroyed the 1919 settlement but redrew the map of Europe in a way about which Germans had only dreamt for centuries, and attained at relatively little cost, had an irresistible attraction.

A people at war, especially when (as Germany was from 1942 to 1945) it is fighting for survival against overwhelming odds, finds it difficult to distinguish between the interests of the state and nation and those of the regime. Even if they were able to make this distinction, it was not countenanced by Germany's enemies, at least after they had committed themselves to the policy of unconditional surrender. As these enemies closed in on Germany, the German people at war, especially on the eastern front, believed themselves to be fighting, and were in fact fighting, not simply for a particular regime but to save German land, German property, and German people from the despoliation which they feared defeat would bring, and which indeed it did bring.

It is necessary finally, in passing judgement on the German people in the Nazi period, to avoid the danger of pharisaism. The war crimes, crimes against the peace, and crimes against humanity of the victorious Allies, whether or not they were comparable with those of Nazi Germany, have not been probed, punished, or publicized as those of Germany and the other vanquished powers have been. The United States, the British Commonwealth, and the Soviet Union bear their own burdens of guilt accumulated in the course of the conflict, and even in relation to the perpetration of the Holocaust they are not without some measure of responsibility. The British people prefer to believe and are entitled to hope that subject to the same stresses they would not have been capable of some of the actions of Nazi Germany, but we have no assurance that this is so. We are inclined, even while distinguishing between the Nazi regime and the German people, to attribute to the latter a measure of collective guilt, but the concept of collective guilt is one we question when (as in relation to the alleged guilt of the former European imperial powers) it is applied to ourselves. The victory of the United Nations has determined the perspective from which the history of the Second World War has been written, the perspective of German historians no less than of others. The Second World War is widely regarded today as a moral crusade fought by the Allies on behalf of human rights and the self-determination of peoples, but while this view of the conflict is by no means wholly false and the victory of the Third Reich would have been a setback for civilized values of unimaginable proportions, the war was also fought on behalf of traditional national interests and was the last in a series of struggles waged to prevent the hegemony of a single power in Europe.

Nevertheless, it is a fact that the great majority of the German people accepted the Nazi regime at least passively, where they were not active supporters of it. Those who chose the path of resistance were a tiny minority; neither the armed forces (inhibited as they were by the German officers' personal oath of loyalty to Hitler), nor the churches (inclined as they were to view Nazi Germany as the bulwark of Christendom against the east), nor the bureaucracy made any collective stand against the regime, even if these institutions did jostle with the Nazi party to preserve their position in the national life, and individuals and small groups within them provided the main source of the resistance that did develop. As Karl-Dietrich Bracher notes, the tradition of humanity and enlightenment, grown so weak since 1848, could not be revived to meet the challenge, and this is a burden which the German people today must bear.

IV

The German resistance has not been celebrated as the resistance movements in the countries of German-occupied Europe have been. Those who took part in it were resisting not a foreign invader but the government of their own country, for the last six years in time of war, and the implication that their activities were treasonable has been an obstacle to the full recognition of their contribution in the Federal Republic of Germany to this day. The German Democratic Republic sees itself as the heir of the German Left crushed at the outset of Nazi rule, and has not until recently sought to honour the resistance carried on from within the German military and bureaucratic establishment.[2] Britain and its allies never sought to encourage the German resistance, or to enter into the same sort of relationships with it that were established with resistance forces in the occupied countries, either before or during the war; its significance was doubted and its aims were regarded as suspect; its best-known action, the July 1944 plot, came only at a stage when Germany was clearly losing the war and was open to the interpretation that, as A. J. P. Taylor so brutally put it, the conspirators 'were more anxious to save Germany from the Russians than to save Europe from the Germans'.[3] The conservative resistance repudiated Nazi barbarity and sought to return to the traditions of civilized behaviour but it still embraced German national aims unacceptable to Germany's opponents. There may even be doubts as to how far the various acts of individual and group opposition within the Third Reich, lacking as they were in effective co-ordination and uninformed by any agreed common doctrine or programme, are properly called a resistance at all.

Various of the contributions of this book shed a more favourable light on the German resistance. Klemens von Klemperer shows that the contrast between the *Widerstand* in Germany and the resistance movements in other European countries was not as great as has been supposed. The opponents of Nazi rule in Germany, moreover, had to face obstacles (more effective Nazi control, lack of popular sympathy, the suspicion of treason) which its opponents in other countries did not have to surmount. David Astor speaks of the gulf that separates those who live in a modern police-state from those who do not, and the effort of imagination that is required to grasp how narrow the options of those who seek to resist authority in the former may be. Karl-Dietrich Bracher writes that the decisions of those who chose resistance had often to be taken in loneliness and without the prospect of public recognition of their sacrifice or of a halo of glory. Marion Gräfin Dönhoff in a lecture on 'The Ethics of Resistance' in the series, the text of which is not reproduced here, set out the contrast between resistance in a constitutional democracy such as the Federal Republic today, and resistance in an authoritarian state, and spoke of the schizophrenia or loss of identity suffered by those in Nazi Germany who led a double life as servants of the state and secret agents of the resistance movement.

[2] East German historians today argue that Stauffenberg and his circle represented the 'progressive' wing within the conservative resistance, and that the attempt of 20 July was patriotic and anti-fascist. See *Süddeutsche Zeitung*, 9 July 1984.

[3] A. J. P. Taylor, *The Course of German History*, (London, 1945), 262.

David Astor sets out the case, disputed in the chapter by Peter Ludlow, against the British government's failure to respond to overtures from resistance groups inside Germany, and argues in particular that an appropriate response at the time of the Czech crisis in 1938 might have led to the overthrow of Hitler and the avoidance of the war. This raises some large questions. Can a great power at the height of an international crisis prudently engage in the subversion of its adversary? What assurance did the British government have in 1938 that a coup against Hitler would be successful, that a new German government of the army rather than of the Nazi party would be an acceptable partner in Europe, or that the conspirators rejected Hitler's programme of military expansion rather than merely disagreeing with its timing? After the entry of Russia and America into the war did Britain, as a partner in the Grand Alliance, have any scope for independent action in this area? Most basically, to reopen the matter of relations between Britain, or the Allies as a whole, and the German resistance, is to raise again the fundamental question as to what were the issues at stake in the war itself.

The proposition that Britain and its allies were at fault in their sometimes callous attitude to resistance groups in Germany draws much of its strength from the idea that the chief issue in the struggle was the removal of Hitler's rule and the scourge of Nazism from Europe. If this was indeed the chief object of the struggle then anti-Nazi groups in Germany might have been expected to be able to help bring it about, as the German revolution of 1918 had brought an end to the rule of the Kaiser. The anti-German feeling generated in Britain and elsewhere during the war, on this view, was not only ugly and discreditable in itself but was an obstacle to the pursuit of the Allies' own purposes. One may understand the sense of betrayal felt by anti-Nazis in Germany who looked in vain to the allies for signs of encouragement. Marion Gräfin Dönhoff, in the lecture to which I have referred, spoke as follows: 'I was with Peter Yorck[4] in Berlin in January 1943, the day on which the Allies announced unconditional surrender as the conditions for the capitulation. When we heard this news on the radio we said with one breath: "Now it's all up with us." For unconditional surrender meant that Nazis and anti-Nazis would combine and do their utmost to prevent just this. For the opposition the great difficulty had been the right timing. As long as Hitler was victorious on all battlefields, any attempt to eliminate him would have created a new stab-in-the-back legend. One had, therefore,

[4] Peter Graf Yorck von Wartenburg, a member of the Kreisau Circle and descendant of the great general of the Napoleonic era.

to wait until the tide turned. But one could not risk waiting until the Allies had no more interest in negotiations with the Resistance because they were anyway within a short distance of victory. This was why it was important to make a distinction between Nazis and Germans.'

Today the notion that the Second World War was an anti-Nazi or anti-fascist war serves important political funtions: it casts the victors in an agreeable light; it papers over the differences that even then existed among them about the shaping of the postwar world; it helps reconcile Germans to non-Germans, the Federal Republic of Germany to NATO and the European Community, the German Democratic Republic to the Soviet Union and the Warsaw Pact; it sets aside those historic issues that have divided Germany from its neighbours and made the German problem in one or another of its forms a central issue of international politics in Europe throughout modern times.

In fact the Second World War was not only or even chiefly a war against Nazism, either for the Western allies or for the Soviet Union. Like the First World War for the Allied and Associated Powers, it was fought to prevent German domination of Europe. The fact that the particular German government against which they were fighting was a Nazi one did of course profoundly affect the nature of the conflict. It meant that the Allies in their war-propaganda ranged themselves against Nazi principles as well as against German expansionism, just as in the First World War, after the collapse of imperial Russia, they ranged themselves against authoritarianism and 'Prussianism'. The barbarities of Nazi rule in Germany and occupied Europe quite genuinely shocked the Allies, while also creating a tidal wave of antagonism towards the Nazi regime which they were able to mobilize for their own purposes. The revelation in 1945 of the full extent of Nazi atrocities magnified this antagonism and made the war seem in retrospect to have been more of a struggle for human rights than in fact it had been. If Germany had not had a Nazi government in the first place, or if this government had been successfully overthrown from within, the so-called German problem, the problem of fitting Germany into the European political system without endangering the independence of other states and the self-determination of other peoples, would have been easier to solve than it was. But it would still have been there, just as in the aftermath of the Nazi era we have still to live with it today.

If in fact there was no complete identity of interest between Britain and its allies on the one hand and the German conspirators on the other, this in no way diminishes the achievement of the latter. It is to the credit of those who strove to find a new path for Germany that they were not mere ciphers of British or Allied policy, but espoused objectives that were rooted in the traditions of their own country and took account of the interests and aspirations of the German people at the time. Their struggle against the Nazi way, had it succeeded, would not in itself have disposed of the conflicts of interest and objective between Germany and the Allies, but it might at least have established the conditions in which reasonable adjustments and compromises were possible.

V

The response of Great Britain to the external challenge of the Third Reich, the policy of appeasement, has come to be regarded as one of the classic errors of foreign policy in modern times, an object-lesson in how not to deal with authoritarian governments, to be cited and applied in relation to Sir Anthony Eden's policy towards Nasser's Egypt, President Johnson's policy towards China and North Vietnam, President Reagan's policy towards the Soviet Union, and many other cases to which it has little if any relevance.

Appeasement, the policy of seeking to satisfy a dissatisfied power through concessions to its demands, can take different forms. We may distinguish between a policy of unilateral concessions to the power demanding change, and the pursuit of agreements in which concessions are made on both sides. Appeasement may be passive, as when Great Britain and France failed to react to Hitler's remilitarization of the Rhineland, or it may be active, as when at the Munich Conference they not merely failed to prevent the dismemberment of Czechoslovakia but put their own imprimatur upon it. Appeasement of a dissatisfied power carried out from a position of strength, such as the British government enjoyed during the period of the Weimar Republic, may be very different from appeasement based upon weakness. What is conceded in a policy of appeasement may be an outworn and unjust privilege, or it may be a vital principle. Any attempt to determine whether appeasement is a sound or an unsound policy for the Western democracies to adopt towards their adversaries will turn on these distinctions and others.

Some form of appeasement or accommodation was an essential element in any British policy of dealing with Germany in the inter-war period and had already entered the minds of British policy-makers by the time of the Paris Peace Conference. Germany's weakness in 1919 was inherently temporary; it was bound in the course of time to recover something of its power and prosperity. Given this, it was in the interest of the Western powers that Germany should resume its place in the international polity and economy; these were the objectives for which men like Ramsay MacDonald and Austen Chamberlain worked in the 1920s; they resulted in the Locarno treaties, reconciliation between France and Germany, alleviation of the reparations problem, and the restoration of German prosperity which, as Keynes had argued in *The Economic Consequences of the Peace*, and as the policies adopted by the Allies towards Germany after the Second World War later confirmed, was to the benefit of all Europe.

The restoration of Germany to the position it had occupied before the Great War implied that it would become once again the naturally dominant power on the European continent. But despite its supposed instinctive opposition to any threat to the European balance, Great Britain had shown that it was able to live with the kind of primacy in European affairs which Germany enjoyed in the Bismarck era; it was not until Wilhelmine Germany became dissatisfied with hegemony in Europe and sought to challenge Britain's naval and world position that fundamental conflict arose. It was a theme of Hitler's, as of many Germans' before him, that the central issue in international politics was the balance of power not in Europe but in the world as a whole; Britain had sought to maintain the European balance only so as to promote its own dominance in the world outside Europe, where by the late nineteenth century the British Empire, the United States, and the Russian colossus jostled for position, and Germany could compete effectively only if it united Europe under its own leadership. Hitler's idea that there was no inherent conflict between the British Empire and a Germany whose efforts were confined to the achievement of hegemony in Europe had a powerful appeal to those in Britain who placed imperial interests first. Such a German hegemony implied that the rights of the small states created or enlarged by the 1919 settlement would be curtailed or extinguished; but these states enjoyed effective sovereignty in the post-1919 period only because Germany and Russia had temporarily ceased to be great powers; if we are to say that the liberties of small nations are sacrosanct, this is a principle which Britain was not able to make viable in eastern Europe in the 1930s, any more than it is able to do so today.

The German claim for treaty revision in the inter-war period was rooted in a demand for justice, especially for just application of the principle of national self-determination, and sympathy for this demand in Britain provided one of the chief rationales for the policy of appeasement, sometimes sincerely believed, sometimes a pretext for the avoidance of Britain's responsibilities. The German case for just change, stated with moderation by Stresemann and other leaders of the Weimar period, and crudely and hypocritically by Hitler, did represent a powerful and in some measure a valid critique of the international posture of Great Britain and France; the Versailles powers purported to stand for international law and the peaceful settlement of disputes, but these principles protected a distribution of territory that they had just brought about by their victory in war. The force of this argument was recognized in Britain by those who addressed themselves to the problem of peaceful change; Article 19 of the League Covenant provided for treaty revision recommended by the Assembly; if in the past war had been the main agent of change in international relations, then the abolition of war had to be accompanied by a new willingness to accept change by peaceful means. But as E. H. Carr wrote in his famous defence of the Munich settlement, peaceful change does not come about simply by virtue of recognition of the need for justice; there has also to be the element of application of power. The agreements concluded in Munich embodied both elements, recognizing both the justice of Germany's claims and the shift of power towards Germany that had taken place.[5]

But Neville Chamberlain's policy of appeasement certainly failed, at all events if we assume that its object was the declared one of a lasting understanding with Nazi Germany, rather than that of postponing an inevitable war until Britain was ready to fight it. Hitler was not Bismarck, and however Britain might have reacted to a German primacy restored under the leaders of the Weimar Republic, the outright German domination of Europe that Hitler envisaged could not have been compatible with British interests. The policy of accommodation of Germany, which earlier British governments had pursued from a position of strength, in Chamberlain's time was a reflection of weakness, and had the appearance of mere capitulation to bullying and threats. The argument put forward for the justice of Nazi Germany's demands for change, even if we disregard the fact that Hitler himself treated notions of justice in international relations with contempt, as he did the principle of national self-determination, contains a basic flaw: justice can have no meaning in international relations except in a context of order. Hitler's policies did not merely challenge the international political *status quo*, they brought the international system itself into question; when he

[5] E. H. Carr, *The Twenty Years' Crisis 1919–1939* (London, 1939).

overran France in 1940 he said, truthfully enough, that what he had overthrown was not the Treaty of Versailles, but the Treaties of Westphalia. The sympathy in Britain for the justice of Germany's claims was, as Martin Wight has written, a 'subversive sentimentality'.[6]

It is clear in retrospect that Britain could have appeased or accommodated Germany without overturning the international order only in the context of a structure of alliances. Peter Ludlow argues below, surely correctly, that the ultimate failure of British policy in the 1930s lay in the nationalist and unilateralist premises on which it was based. Germany was defeated in 1914–18 by a combination in which Britain was linked at least for part of the period not only to France but to Italy, Japan, the United States, and Russia. The alliances that Britain needed in order to maintain a balance against a revived Germany she did not get until Hitler, by attacking Russia and declaring war on the United States, provided her with them. It is true that it was not only the British in this period whose thinking was unilateralist. Alliance with the United States was never available to Britain in the inter-war period; in the case of the other ex-partners, however, it might have been available had Britain been ready to pay the price: for France, a firm commitment to French security; for Japan, hegemony in East Asia and a loosening of British connections with the United States; for Italy, change in the Mediterranean and Africa; for Russia, acquiescence in Russian ambitions in the Baltic and eastern Europe. Britain's reasons for being reluctant to pay the price for these alliances were understandable, in some cases even honourable and creditable; but her range of choice in foreign policy was narrower than British leaders or the British public then supposed.

Britain's failure to respond effectively to the challenge of the Third Reich was rooted in an intellectual failure to comprehend the world of international relations in the 1930s as it really was, but also in a failure of courage and resolve in facing up to the limited and indeed brutal choices which were the only ones she had in the situation which she had allowed to develop.

VI

We are inclined to regard the Third Reich as an aberration in the history of the twentieth century, a reversion to barbarism which briefly stalked the earth but whose crushing defeat in 1945 led to a new beginning,

[6] M. Wight, *Power Politics* (London, 1979), 203.

much more decisive than that of 1918 or 1815. Hitler, unlike Napoleon, is said to have left no monuments. The United Nations Charter, which supplies the common language of world politics today, even if it does not determine its conduct, is the work of the coalition that destroyed Nazi Germany, and for all that divides them, Western liberalism and Soviet communism, both of them heirs of the European enlightenment, are united in their abhorrence of Nazi concepts and doctrine. The view that the Third Reich was the creation of one man of demonic powers and epochal ambitions, who by a unique combination of circumstances was given the opportunity to implement his sinister design, tempts us to see the Nazi phenomenon as a singular occurrence. The theory that Nazism was the special product of German history, the culmination of racial myths, *Volk*-worship, ideas of heroic leadership, preoccupation with death and violence, and dreams of *Reich* with deep roots in the German past (an idea propounded by some British writers during and after the Second World War, but also by apologetic German historians), serves to confine the significance of Nazi ideas to the experience of one nation, which has now apparently repudiated them utterly.[7] Even if we take the much more pessimistic view that the Nazi phenomenon was part of a convulsion of Western civilization as a whole, the enactment of a suicidal impulse long evident in European culture and perhaps conjured up by Western civilization itself, we might still believe that the convulsion had run its course by 1945. We are reluctant to integrate the Third Reich into the history of this century, to acknowledge that it provides any clues to the understanding of more recent events.

The question is whether despite its defeat Nazi Germany was not after all part of the mainstream of events in the twentieth century. Along with its atavistic elements the Third Reich contained, as chapters in the present volume show, elements that may be judged to be forward-looking or progressive, in the sense that they involved changes that have since been generalized and taken further: a concern with social equality and mobility (although not including women), eradication of social obstacles to development, promotion of a secular outlook, the conquest of unemployment, attention to the impact of development on environment. The regime that gave us the Volkswagen and the Autobahn also made its signal contributions to the theory and practice of one-party rule, political education through youth movements, the use of propaganda,

[7] See e.g. R. d'O. Butler, *The Roots of National Socialism 1783–1933* (London, 1941), and A. J. P. Taylor, *op. cit.*

the secret police and the concentration camp, and other aspects of the apparatus of modern states in the era of mass political awareness that flourish widely today.

Even the self-consciously backward-looking elements in Nazism—the exaltation of emotion over reason, the cult of violence, the repudiation of decadent civilization and embracing of a barbarian past—are not out of step with the reactions of other societies in the course of this century to economic development in the industrial era, but are standard responses, aspects of a process of adjustment which are to be observed in all parts of the world today, even if they manifest themselves in different ways. The Nazi formula that satisfied not only the aspirations of the masses for social and economic advance but also the longing for expression of national identity and for *Volksgemeinschaft* was basic to the regime's public appeal. We should not fail to notice that the combination of nationalism and socialism provides the prevailing ideology in the greater part of the world today.

It is true that the direct influence of the Third Reich on later events has been via the reactions that were provoked against it: championship of human rights by the Western powers in the early post-war period; the movement for European unity; the foundation of the state of Israel; the disappearance from polite international discussion of Hitler's language of survival of the fittest, racial hierarchy, and the domination of inferior peoples. Even in the countries of the Third World, so much less affected by the Nazi experience than those of Europe, Russia, and North America, there is some reluctance openly to embrace Nazi precedents and models. But as the history of the twentieth century has unfolded since 1945 and the record of government by terror, oppression of minorities, war-crimes, and genocide has built up, the idea that the Third Reich was abnormal or exceptional in its bestiality has become more difficult to sustain.

2

ADAM VON TROTT: A PERSONAL VIEW

DAVID ASTOR

I FIRST met Adam von Trott in the Porter's Lodge of Balliol in September 1931: we were signing on with the other new arrivals.

I had just been spending a term at Heidelberg in what were to be the last eighteen months of the Weimar Republic. During one week-end of that summer, a few thousand members of the Nazi Party arrived to hold a National Rally in our little city. Very few people, at that time, except the Nazis themselves, thought they would come to power. Yet this glimpse of a mass of ordinary-looking men determined to bring back militarism was highly alarming.

All this made me curious to get to know the German Rhodes scholars at Oxford. Two of them became my friends. They were both to prove themselves among the most independent and courageous people of our generation. One was Adam von Trott: the other, Fritz Schumacher (who later wrote *Small is Beautiful*).

These two were themselves friends; both came from liberal-conservative families, one diplomatic, the other academic. Of these two families, the Trotts were much the more free-thinking. Theirs was the kind of almost eccentric home where every question of the day was debated, by the elder brother from the angle of a Marxist, by Adam from that of a Social Democrat.

With this background, Trott found the atmosphere of Oxford instantly congenial. He soon became a popular figure and a friend of some of the ablest younger dons of that time, such as Isaiah Berlin, Dick Crossman, Maurice Bowra.[1]

Then in January 1933, came the event that was to change everything: Hitler, having won enough votes in the previous elections, was able to bully his way to power. Almost at once, the situation of the German students in Oxford became subtly different. The boldest no longer spoke out against Hitler in public debate, as Trott had done so vigorously

[1] Sir Isaiah Berlin (1909–   ), liberal philosopher, Lecturer in Philosophy, New College, 1932, Chichele Professor of Social and Political Theory 1957–67); Richard Crossman (1907–70), Labour politician, Fellow of New College 1930–7, Lord President of the Council 1966–70; Sir Maurice Bowra (1898–1971), scholar and wit, Fellow of Wadham 1922–38, Warden 1938–70. (Ed.)

before. From that moment and for the years ahead, there was to be an inevitably growing gap—in experience and in understanding—between them and the rest of us.

What were the feelings of these German students? They must have known some unfamiliar sensations—such as fear, shame, utter incredulity. And there was, moreover, virtually nothing that a student, or anyone else, could do about the situation. The main characteristic of the new regime was that it allowed no opposition whatsoever. The Communist and Socialist leaders who had said they would oppose were simply rounded up and made to disappear.

Caught in this situation, Trott's first reaction, as I recall, was gloom, tempered by challenge. I remember a walk in the country when I told him that if he defied the Nazis they wouldn't answer him with arguments. I tripped him up and pushed him to the ground to suggest what they would do. It was a silly episode, but it shows at least what we were talking about: his belief that he could maintain some sort of opposition.

He decided to spend the next years completing his law studies in Germany, and also meeting his fellow Germans of all sorts to discover their attitude. He drew most comfort from Berlin working-class reactions. He talked to friends at this time of not giving the country over to Hitler.

He also spoke of the hatred that the rest of the world felt for Germany during the Great War and feared that Hitler would inevitably bring it back. He yearned to retain some sense of common humanity between nations and hoped that Germans could be seen as a part of the normal world, not as a uniquely criminal people.

With Hitler talking of a Master Race and behaving with the utmost brutality, it was difficult to maintain this line of thought. Yet Trott, to the annoyance of some British friends, continued to think and speak as if a normal Germany still existed within the Third Reich. He did this because he thought that *not* to do so was to play Hitler's game. He believed that to accept Hitler's monolithic Third Reich was fatal for Germany and utterly against the interests of Germany's neighbours.

After some three years of his sub-life in Germany, Trott began to feel compelled to interfere in the course of events. Although it may seem exaggerated in a man still in his twenties, he felt a personal obligation to try to stop the Nazis from taking Germany in so terrible a direction as making war. By the end of 1936, he had accepted that there was no possibility of popular action in Germany. He therefore conceived the plan of making a visit to the Far East and the United States. His purpose was partly to look at Germany in its world context: for example, he toyed with the idea of the Anglo-Saxon powers drawing off, at least temporarily, the energies of Nazi Germany in an economic partnership in the Far East. But his main concern was to decide what role he personally should play in trying to rid Germany of its perverse regime and in preventing another war.

By this time, he had already heard rumours of discontent among a few individual army officers. Trott did not know many army people: but he early recognized the military as being alone in possessing the means to strike down the regime—but also as being those most inhibited, by their training in obedience, from doing so. Signs of political allies in that quarter were, therefore, exciting and sharpened his fundamental choice. Should he join whatever active opposition might develop and start a life of real danger in his own country? Or should he try to influence events from outside Germany, where he could at least speak out openly?

His journey across the world to think out what to do was much beyond his financial means. How he managed to finance it is striking evidence of how he was regarded by experienced men of the world in Britain. He turned to two older men whom he had got to know well in his Oxford years: both reacted in a way that showed they regarded him as a young man of unusual stature.

One was Philip Kerr (later Lord Lothian), Secretary of the Rhodes Trust. He made Trott a further grant from that Trust. Kerr had himself been an outstandingly able young man. He had been private secretary to the prime minister, Lloyd George, during the war and the Versailles Conference, when little older than Trott was at this time. He had been deeply troubled by the punitive aspects of the Versailles Treaty and this later impelled him to interview Hitler to try to assess their relevance to his thinking. He was made British Ambassador to Washington just before the outbreak of war and was confirmed in that post by Winston Churchill. His loyalty to Trott extended to meeting him in America (no doubt, improperly) during the war.

Another backer was Sir Stafford Cripps, whose son, John, was one of Adam's contemporaries at Balliol. Already a famous barrister, Cripps became a leader of the left wing of Labour, together with Aneurin Bevan. He was an outstanding anti-fascist campaigner and his advocacy of an alignment with the Soviet Union was to make Churchill choose him as his ambassador to Moscow. Cripps helped finance Trott's journey with his own money.

With their introductions, Trott made friends in the States with people like Roger Baldwin, a great fighter for civil liberties; Edward Carter, of the Institute for Pacific Relations; and Reinhold Niebuhr, the eminent theologian and writer, who was to defend Trott most resolutely when his name later came under attack.

It was inevitable that Trott should come under American and British suspicion during this visit to the United States and China. The reason for this was that he was pursuing two courses that appeared incompatible. He was meeting influential people as an anti-Nazi: but he was also taking precautions to keep in with Germany's official representatives. He was absolutely determined to keep open his option to return to Germany, in order to join the embryonic opposition there. This meant being sufficiently approved by the German authorities to get employment in, for instance, the Foreign Ministry: it might ultimately mean being able to join the Party. Obviously, none of this would be possible if he had become suspect to German diplomatic officials while abroad. Indeed, to gain their confidence, it was essential that he should make occasional visits to German embassies wherever he was. Such visits were, however, easy to detect, if he was followed by British or American security officers. And this is what happened. Those visits were later to be considered by the American and British authorities to be conclusive evidence against him.

Meanwhile, Trott was often tempted not to return to Germany: he spoke of seeking an academic post in America. Indeed, when Hitler seemed about to bring Europe to war in 1938 at Munich, Trott became so filled with foreboding that war was now inevitable that he spoke then, just before returning to Germany, of settling in America.

The decision he took to travel back to Germany to engage in treason against Hitler is not one that can be easily imagined. He took it while alone. And he realized that, once taken, it would be irresponsible to tell friends what he had decided to do—unless they were themselves already involved or especially able to help. To whisper to his Oxford friends, for instance, that what he would appear to be doing in Germany was not what he would really be doing would be merely frivolous: he would be endangering his whole purpose and the lives of his fellow conspirators merely to gain personal approval.

He chose to tell me of his decision (originally in a letter from China) because he needed my help. He knew my parents and thought that they, as supporters of Chamberlain's government, were his best hope of making contact with that government. Cripps had no such possibility.

And it was to be Trott's hope to bring about a working partnership between London and the Opposition in Germany.

He arrived back in Germany from the Far East in December 1938 and, before that month was ended, had made an astounding discovery. Through a friend in the Foreign Ministry, he learned that, behind that establishment's punctilious façade, there had been activities of a kind of which he had scarcely dared to dream. Clandestine acts of the most promising kind had been taken by no less a person than the head of the Foreign Ministry staff, the ultra-professional Ernst Freiherr von Weizsäcker himself.

This had happened some three months previously when Hitler was threatening to invade the Sudetenland province of Czechoslovakia and there had seemed every probability of war. In that circumstance, senior commanders in the German Army, Beck, Halder and others, had secretly decided to arrest Hitler at the moment that he gave the order to cross the Czech frontier. Their preparations are said to have included the movement of an armoured unit to Thuringia to cut off the SS stationed in Munich from reinforcing, via the new *Autobahn*, the SS in Berlin. The mutinous generals were confident that most of the German public would be greatly relieved to be spared another war and would support them.

The most extraordinary step that they took was to ask Weizsäcker to inform the British Government of their intention. This meant to work with a foreign government against their own government. With the head official of the Foreign Ministry willing to deliver their message, the chief officers of the German State, both civil and military, were preparing a coup against their government.

The transmission of their message had to be kept absolutely secret. It had to reach the British Government without passing through the German Ambassador in London. So, who should convey it? The British were not likely to receive a German diplomat who was not introduced by his own embassy.

The method used was highly ingenious. There were two brothers serving in the German diplomatic service at that time, Theodor and Erich Kordt, whom Weizsäcker trusted completely. Theodor was Counsellor at the London Embassy. Weizssäcker sent for Erich. He let him into the clandestine plans of the disaffected officers. He then asked him to make a holiday visit to London, ostensibly to talk family business with his brother, but in reality to convey the outline of these plans by word of mouth to his brother, the Counsellor. The latter should then ask that he might have an exceptional and strictly private meeting with Lord hat is to say, a meeting unknown to his own ambassador. In indication of the planned coup, Theodor Kordt was to advise ifax that the British Government could be most helpful simply ıng firm against Hitler's demands.

This meeting took place as planned: the Counsellor to the German Embassy was even admitted by the garden entrance to 10 Downing Street to avoid possible observation. But, as we all know, the British Government decided to disregard his message. They proceeded to go ahead with the Munich Agreement. That agreement effectively gave the Sudetenland to Hitler. He was, therefore, able to send his army into Czechoslovakia without any risk of war. That the German public had, indeed, been dreading another war, as the generals believed, was amply demonstrated in the emotional scenes that greeted Chamberlain, not Hitler, in the streets of Munich when the agreement was announced. (Chamberlain was to receive shoals of grateful *German* letters for days thereafter.) But all those involved in the plot to arrest Hitler must have been utterly disgusted and downcast. The ideal and perhaps the only opportunity to strike Hitler down in circumstances that would have been understood and welcomed by the German public had been thrown away.[2]

No one seems to know what happened at the British end of this tragic story—tragic because this was the most formidable trap ever to be laid for Hitler. Did Halifax doubt the authenticity of the eminently respectable Counsellor, Dr Kordt? Perhaps the message was too extraordinary to be accepted without confirmation. In that case, why were no further enquiries made, for instance by sending someone to talk equally discreetly to one of the senior officers indicated by Kordt? Was it, perhaps, simply impossible for the British Cabinet to imagine that Hitler's chief civil and military officers really wished to betray him? Or did the Cabinet just prefer their own way of dealing with Hitler? We don't know.

It is sometimes now suggested that Britain, already at that time, had long-term aims, such as permanently reducing the national strength of Germany, which could not have been achieved if the generals' coup had been encouraged. But can it be seriously believed that deeply pacific men like Chamberlain and Halifax had geopolitical aims at that time of

[2] On these events see *The Memoirs of Ernst von Weizsäcker, Head of the German Foreign Office 1938–1943*, tr. J. Andrews (London, 1951); Erich Kordt's memoirs, *Nicht aus den Akten* (Stuttgart, 1950); Hans Rothfels, *The German Opposition to Hitler: An Assessment*, tr. L. Wilson (London, 1961, repr. 1970).

weakness and anxiety, plans which were scarcely imaginable until Britain had both the United States and the Soviet Union as allies? And is it imaginable that they (or anyone else) could have preferred an immeasurably destructive world war to a putsch by his own generals as a way of dealing with Hitler? Halifax's alleged remark to Dr Kordt at a subsequent informal meeting, that the British Government was too far advanced with its own plans to be willing to alter them, sounds more likely.

Trott may not have been told the full details of the Beck–Weizsäcker plan—which, after it miscarried, must have been kept more secret than any other secret in Berlin; but he certainly knew its essence. He was excited to learn that senior army chiefs had been so much opposed to fighting what they called 'Hitler's war' that they were ready to plot with a foreign government. He was fully determined to work for a second such attempt and at once offered to use his connections in London for that purpose.

He was, however, soon made aware that a second such attempt could not be laid on immediately. Hitler's greatest bloodless victory had removed, at least temporarily, the possibility of arresting him as plainly a madman who was about to plunge Germany into war and restored his reputation as a wonder-worker. There were also immense practical difficulties. The generals could not control where they might be posted or when they might be replaced. As neither telephone-calls nor written messages were safe in Nazi Germany, personal visits had to be made to prepare any clandestine plans; nothing on a large scale could be improvised suddenly.

Trott's offer to help was, however, accepted: but his mission was never to be as clear-cut as that of Dr Kordt. The German Opposition's hopes of preventing Hitler from starting another war were now more slender and depended on less likely possibilities. They were to find themselves asking London both for *less* appeasement and for *more*: for stiffer military opposition, to warn Hitler off thinking he could attack Poland with impunity, and, simultaneously, for Hitler to be distracted from war by some—almost any—negotiations.

Trott was given sufficient contact with Weizsäcker to co-ordinate his efforts with those of others: Erich Kordt, for instance, was to be sent back to London once again. Trott's mission was to explore means of keeping Hitler talking: but he was quite aware of the secret urging that Hitler be simultaneously put under a greater military threat.

His first visit to England after his return from the Far East was in

February 1939 and its purpose was the modest one of sounding out the state of British opinion. But even that had its pitfalls at this time of mounting tension. It came as a surprise to Trott to discover the strength of feeling among his liberal-minded British friends towards the very idea of gaining time by political means. 'No further talks with Hitler' had become an absolute imperative among them, since Munich. Trott told me his amazement that they seemed actually to want war. But, for the security of the plotters, he was not able to argue with them that, if peace could somehow be prolonged, a military coup in Germany might be a possibility. He felt a coolness towards himself: however, he judged (rightly or wrongly) that the general British public still wanted peace, even if it meant further talks with Hitler.

Within a fortnight of this visit, Hitler made the move that was basically to change British popular feeling. He ordered his armies to seize Prague and the brutal fraudulence of this act shook the British. The Labour Party ceased to oppose rearmament and the whole country suddenly reached a silent agreement that they must now be ready to face another war. However, this extraordinary transformation was scarcely noticeable to outsiders: the American Ambassador, Joseph Kennedy, for instance, continued to report that Britain would not fight.

The effect of the seizure of Prague on the German opposition was equally dramatic; but it was necessarily different from that on the British, and this further increased their difficulties in understanding each other. The German Opposition could not resolve, like the British, to achieve their purpose through war: their uniting purpose was to prevent a war. Moreover, peace was considered an almost essential precondition for making a successful coup. Once Germany was at war, the popular legend of the 'stab in the back' would work against them.

The military wing of the Opposition now felt sure that Hitler would attack Poland before they could possibly mount another coup against him and that an attack on Poland would—whatever Hitler himself believed—bring on a full-scale European war. The generals therefore decided to try a kind of diplomacy of their own, which was even more unorthodox than the attempt to reach Lord Halifax via Dr Kordt. They sent to London in the summer of 1939 an English-speaking staff officer from the High Command, Colonel Graf von Schwerin. He equipped himself with printed visiting cards giving his Piccadilly address and also his full German military status. Working, naturally, not through his embassy, he used what personal introductions he had been given to try to approach the British Government. He had, however, such difficulty in making contact with the Government, even at a low level, that he turned for help to me, an entirely unqualified twenty-seven-year-old, whose name he had been given by a Hamburg businessman, Wilhelm Roloff, whom I had met through an émigré friend, Erwin Schueller. I consulted Trott: he said Schwerin was not a Nazi, although not necessarily an anti-Nazi, and that I should certainly meet him. He proved to be a genial, blunt individual who spoke his lines unhesitatingly.

The message he wanted to bring to the British Government was simple and precise. The High Command expected an attack on Poland that summer: Hitler did *not* believe that Britain and France would go to war over Poland; it was useless for the High Command to tell him that they would, as they had predicted war on previous occasions, particularly over Czechoslovakia, and had always been wrong. If Britain this time really meant to go to war, there was surely no advantage to Britain in keeping Hitler in any doubt. But the only way to convince him was by actions. The British should send a warship into the Baltic to exercise off Danzig (the area under dispute) without delay. They should transfer RAF squadrons to French airfields facing the German border. And they should replace the British Ambassador in Berlin by a military man. 'He will scream, but at least he will believe that you mean war—if you do! You should also invite some friend of Göring, such as Milch, to see for himself that you really have a serious aircraft industry—they only believe their own cronies.' Those, near enough, were Schwerin's undiplomatic words.

When I approached my only contact in the War Office, Robert Laycock, and, through him, was passed to a senior officer engaged in German Intelligence, I was surprised to be told in the grandest of manners that he regarded Schwerin's visit to London, 'at a time when relations between his country and ours are as bad as they are, as a bloody cheek'! No further discussion was proposed. I therefore had to leave Schwerin to his own devices—he had other contacts, but they did not reach higher than the Chief of Naval Intelligence—and some weeks later received a thank-you postcard from him when he returned to Berlin. He did me one personal favour, which confirmed his *bona fides*: I asked him whether it would still be all right for me to visit Berlin in July; that is to say, I asked him the approximate date of the expected war. He told me (with only a moment of professional embarrassment) that July would be quite all right.

Meanwhile, a last clandestine effort was being attempted from the Foreign Ministry and this directly concerned Trott. Believing that Hitler still preferred bloodless victories, a scenario was evolved in the hope it might draw him away from his attack on Poland. (This was before the Ribbentrop–Molotov Pact, which was to make such possibilities hopeless.) The idea was to find a way of suggesting to Hitler that he could recapture the position he had held at Munich as 'arbiter of Europe' and, at the same time, gain his 'Germanic' aims of annexing Danzig and its Corridor. The idea was that he should astonish the world and restore to Czechoslovakia its political freedom—with, of course, a proviso that it must remain disarmed. After such a gesture, he could expect to take the Danzig Corridor by negotiation. But who was to make this near-fantastic proposal to Hitler? The German plotters thought that it could only come from Britain.

The authors of this ploy (it is not known who devised it or how large a part Trott played in its invention) were certainly not imagining that this device would form the basis of a peacc scttlement. It was conceived as a possible way of enticing Hitler into essentially meaningless negotiations—which might gain time for a second attempted military coup. This was plainly the understanding Trott gave me when he came to London to try to sell the idea.

He came in early June but this time he had to take precautions. If he met British ministers, this might become a matter of curiosity to the German Embassy, as he did not have official authorization to have such talks. He therefore improvised a cover plan. He had scraped acquaintance with one Walter Hewel, a middle-aged businessman of modest intelligence, who happened to be one of the very earliest Nazis and a personal friend of Hitler. Hewel had been given a high-sounding liaison function in the Foreign Ministry. Trott represented himself to Hewel as wanting to use his personal connections in London to promote a better understanding of the Führer. Hewel gave his general approval and thus became Trott's alibi if anything went wrong. To assist this process, Trott asked me to meet Hewel over drinks in a Berlin hotel and to build up an impression of Trott's high standing in London, which I did, to the best of my ability.

Trott's final London visit was made in the unimaginably tense atmosphere of an impending war. It is easy to see now, with hindsight, that his mission was foredoomed to failure. The Chamberlain Government could not possibly make any further diplomatic move after the seizing of Prague without risking Britain's hard-won national unity. Trott had been away from Britain too long to realize this. He remained passionately conscious that, once a war had started, it would generate its own terrible momentum. Without being a pacifist, he continued to see war as an incalculable disaster that he must absolutely try to prevent.

He met Halifax, the Foreign Secretary, over dinner at my parents' house and took the chance that was given to him to speak. Halifax was sufficiently impressed to arrange that Trott should have a private talk with the Prime Minister, Neville Chamberlain—in itself an amazing achievement at that stage of events for a young German with no official position.

This talk took place at 10 Downing Street on 7 June with Alec Douglas-Home (then Lord Dunglass), the Prime Minister's Parliamentary Private Secretary, also present. Trott's main purpose was to convey the message that there was, indeed, a German Opposition, well placed to strike at Hitler; but that its chances of success depended greatly on whether war could be averted; he also stressed that this Oppositon would be greatly helped, politically, by British encouragement in any form. His secondary purpose was to fly the kite of the scenario for gaining time through talks on Danzig and Czechoslovakia. That, in outline, was how Trott described the occasion to me.

Later on, back in Berlin, Trott produced a fictitious report for the Foreign Ministry of this and other talks he had during this visit. It was written in Hitlerian prose and was deliberately aimed at reaching Hitler's eye, in the hope that it might make him believe that it would be worth his while to delay a move against Poland. It has, not surprisingly, caused confusion by being taken literally by some historians.

I have a clear recollection of asking Trott on one of the last evenings of this visit to say, in the simplest terms, what he was up to. He put his answer in a joking allegory. You could imagine Hitler as a heavily armed drunkard. His wild behaviour was endangering the lives of his own family and his neighbours. The best way to deal with him might be for two people to take him, one by each arm. One arm would be taken by his strongest relative (the German plotters) and the other by his strongest neighbour (Britain). These two should then persuade him to come for a long walk, with much pretence of helpfulness (negotiations on Danzig, colonies, anything at all). Having got him into a quiet field, they should then hit him on the head with his own revolver (the German Army).

This was, incidentally, typical of his private style of speech. I asked him once how he could possibly imagine that Catholics and Communists could work together within the German Opposition. He compared them to tennis players suddenly threatened by a gunman. The players would co-operate to deal somehow with the gunman and would then return to their game—meaning they had a common interest in restoring the conditions of their own more civilized rivalry.

Trott's talks with Halifax, Chamberlain, and Douglas-Home left a favourable impression on them, according to the latter's interview with Trott's biographer.[3] Trott did not manage to persuade them to attempt the role of partner to the German Opposition in trying to overthrow Hitler; perhaps they did not even grasp what he was inviting them to do. However, he certainly did not alienate them or strike them as unrealistic or dubious in any sense.

This generally favourable reception was in marked contrast to what happened when Trott met his friends in Oxford and London on this same visit. With the exception of the two or three to whom he had confided the full purpose of his visit, he made a lamentable, even a fatal, impression. At this distance, the reason is not hard to see. He was proposing more talks with Hitler, without being able to say that the time thus gained could be used for a military coup in Germany. He appeared to be suggesting further concessions, possibly at the expense of the Poles and Czechs, for no purpose other than to buy Hitler off. He was also unaware that what he was suggesting would have demoralized and disunited Britain.

Trott made one partial exception in his policy of not confiding his real purpose of working for a military coup. He, like many others, had a special respect and admiration for Maurice Bowra, the Warden of Wadham College, a man of great boldness. He therefore took a chance on telling Bowra *half* the truth of his situation. He seems to have told him that he had a connection with the German Foreign Ministry and also to have indicated the complicated relationship that this gave him to those army officers who were prepared to make a coup. Bowra immediately asked what these people would do with Hitler's territorial gains and claims if they succeeded in overthrowing him.

This was a question that greatly worried some of the opposition generals. They did not want to be accused by the German public of giving up what might be regarded as legitimate German claims. At that moment, Hitler was demanding a corridor through Danzig to restore the land-link with East Prussia—quite a popular demand: this obviously presented them with an awkward problem. What was, however, incon-

[3] C. Sykes, *Troubled Loyalty: A Biography of Adam von Trott zu Solz* (London, 1968).

ceivable was that these officers would have wanted to keep conquered Czechoslovakia—when they had but recently been willing to make its conquest the cause of striking Hitler down.

Had Trott put the dilemma facing the plotting soldiers frankly to Bowra and made it plain that they must be thinking more of how they would handle a civil war situation with the SS, rather than how they would settle Germany's frontiers in perpetuity, Bowra would surely have understood and would at least have discussed their problem reasonably. As it was, Trott not fully confiding, conveyed the impression of wanting more appeasement of Hitler in a doubtful cause, so that even his own integrity came into question. Bowra asked him to leave the house. Later he wrote to Felix Frankfurter, Roosevelt's close friend, whom he knew that Trott was planning to see, warning him against Trott in the most hostile terms. This, naturally, wrecked Trott's chances in Washington. Roosevelt was to write to Frankfurter jokingly of Trott as a spy.

At the end of the war and after Trott's death, Bowra remarked to another Oxford wit that Trott was one of the few Nazis to have been hung. Later, he changed his tune: in his memoirs, he told the story of his misunderstanding of Trott and, to his credit, expressed his remorse forthrightly.[4]

The majority of Trott's British friends, honourable and highly intelligent people, took the view, at the outbreak of war, not that he was a spy, but that he was a misguided German 'nationalist'. The term nationalist has, of course, many usages. To be a nationalist in a struggle for independence (say, Greek, Czech, or Polish) is highly respectable. To be a nationalist of an imperial or ex-imperial state (like Britain, Russia, or France) is to be suspect of being an expansionist. But to be called a German nationalist at the time of the Third Reich was a more sinister allegation. It could mean to be the carrier of that sentiment in Germany that had made Hitler's whole career possible.

In the feverish circumstances of that time, it is understandable that Trott's British friends misjudged him. The complexity of his own situation, the extreme emotions excited by the impending war, even an accidental factor like his use of my family (themselves the objects of suspicion, because of the supposed 'Cliveden Set') to reach British ministers, were all enough to provoke accusations in that terrible summer of 1939.

[4] C. M. Bowra; *Memories 1898–1939* (London, 1966).

It is more surprising that some who judged him so severely then still maintain that they were right or, at least, do not openly regret their mistrust of him, as did Bowra. There are those who had an emotional involvement with him and who have written of their relationship as of a lovers' quarrel. Others, such as the permanent or temporary Foreign Office officials who compiled the dossier on him, have inclined to stand by what they did. To admit that they were wrong would mean to be guilty of betraying an honest man, since he was encouraged to keep contacting British embassies in neutral capitals in wartime, at great risk to himself, although there was never any intention to answer his messages.

These ugly confusions took place some forty years ago. They should have been cleared by now. For instance, it should by now be admitted that there was never the slightest suspicion of Trott within the German Opposition itself. It should also be beyond dispute that those British friends in whom Trott confided and who remained his active supporters throughout were at least as good judges of his anti-Nazi purity as those who have smeared him as a nationalist. The man who was his closest political friend in this country and who spoke up for him in Whitehall in mid-war was Stafford Cripps. His relationship to Stafford and Isobel Cripps was almost that of an adopted son. It is simply unimaginable that a man of Cripps's mind and character would have tolerated any trace of German nationalist sentiment in a close friend.

Another intimate friend over many years was Wilfrid Israel. This brilliant young man, who owned family property in Berlin, but was the holder of a British passport, used that safeguard to bring Jewish children out of Germany till the last possible moment. A recent biography of him shows the high regard and close relationship that he and Trott had for each other.[5] Would Wilfrid Israel not have noticed the German nationalism which Trott's Oxford friends thought they detected?

Finally, consider his relationship to Fritz Schumacher, a man of pronounced internationalist sentiment. They finally disagreed over whether to stay in Germany or move to England. Trott felt an obligation to try to save his country—not from its neighbour-states, but from its own regime. Schumacher understood this perfectly well, but thought Trott was assuming too great a responsibility. The differences between them was, I believe, not one of quality, but of category. Trott belonged to the category of patriot.

[5] N. Shepherd; *Wilfrid Israel. German Jewry's Secret Ambassador* (London, 1984).

That term is normally used of a defensive role. He may be defending his country from a domestic enemy, like Wilhelm Tell against the Emperor. In the wars of religion, it was normal for such patriots to call in the help of outside states. Trott and the rest of the German Opposition—whether churchmen or trade unionists or soldiers or lawyers—were patriots of this kind. They were engaged in a civil war, which became an international war. They were divided between those who looked mainly to the West for outside help and those who looked mainly to the Soviet Union. The former were much the better placed internally for taking action and became the main network of Opposition. Trott's personal political position is shown by his recommendation to his fellow conspirators of Pastor Niemöller as Chancellor if their coup succeeded—a figure of strong appeal to Left as well as Right.

What should be obvious is that any Opposition needed membership in those circles able to strike down the Nazis. This meant support in the officer corps. There were soldiers, such as Beck and Oster, who belonged fully to the Opposition. But there were others who had to be won over.

And it was these marginal generals who were most anxious not to be regarded by their countrymen as traitors. They therefore asked most fervently for some assurance from London that if they struck Hitler down they would not be made to appear men who had sold their country. Their hesitations were, obviously, the cause of delay and annoyance within the Opposition.

This ideological gradation between the Opposition proper and some of the military on which it depended seems easy enough to understand. However, nearly forty years after, it is still possible to find scholars who claim they cannot understand it.

Indeed, the judgement of Trott himself seems sometimes still to be stuck in the dark muddle of 1939. For instance, a recent book mainly about Trott by a brilliant but disappointed personal friend, Shiela Grant Duff, repeated all the damaging allegations of 1939, but included little of what has since become known which alters the picture.[6]

Again, consider the high repute of the late Sir John Wheeler-Bennett, our greatest authority on the German Army in that period. His *magnum opus* is a study of the German officers of this period, mainly seen as men

[6] S. G. Duff; *The Parting of the Ways. A Personal Account of the Thirties* (London, 1982).

of no moral courage.[7] Yet it was he who pretended in that book that he did not know Trott, when he had long known him and had greatly helped him in Washington *during wartime*. The late Dr George Bell, Bishop of Chichester, told me that Wheeler-Bennett had also seriously misrepresented his, the Bishop's, dealings with the German Opposition. Dr Bell thought this must be because they did not accord with Wheeler-Bennett's thesis, as he showed no interest in Dr Bell's attempts to correct his impression. Yet Sir John's standing is left unquestioned.

The mission of the Kordt brothers and that of Schwerin have been played down in most British accounts. The lack of British response to them, is passed over, almost without comment. In contrast, the failures or acts of incompetence of the German opponents of Hitler have been played up.

Perhaps the most striking omission is in the reasons given for the cool attitude of the British authorities to the attempted coup of 20 July 1944. It is not admitted that the success of an anti-Nazi coup at that time must have caused London acute embarrassment. The secret Allied policy to partition Germany after the war, with the Soviets controlling the Eastern section, was already settled, at least as a temporary occupation measure. Obviously, no Germans, except conceivably some Communists, could have been expected voluntarily to welcome this dismemberment. So, if any anti-Nazis had overthrown Hitler at that time and offered to talk, this would have caused London confusion.

It is perfectly understandable that nothing was said about this at the time. But the continuing pretence that London would not have wanted to deal with the members of the 20 July coup because of their nationalism, when the truth was that they could not have dealt with them because of the Allied intention to partition Germany, is surely propaganda, not history.

Many British historians of the last war were part of the military or propaganda apparatus of that war. It is, therefore, not altogether surprising that they may seem to be unconsciously justifying basic British wartime policies and accepting the then prevailing attitude that everything German, including the Opposition to Hitler, was deeply suspect. It is more surprising that the younger generation of historians, both British and German, do not seem drawn to revising some of the more debatable judgements of that time.

[7] J. W. Wheeler-Bennett; *The Nemesis of Power: The German Army in Politics 1918–1945* (London and New York, 1954).

Two factors may, perhaps, contribute to this apparent lack of interest. They concern the general character of modern dictatorships and our difficulty in understanding them. First, it is the case that any thorough-going modern dictatorship, whether of the left or right, can only be disrupted internally from above, i.e. from the upper echelons of their own society. We have seen this in Portugal, Spain, Greece—and in the Soviet Union and China, where the only major changes have been brought about from inside the government apparatus itself. This means that the internal transformation of modern totalitarian regimes can only be brought about with the help of their soldiers and top officials—or even with that of their secret services, their police chiefs, and such like. No matter who instigates a political change, the mechanics of bringing it about can only be operated with the help of such people. Now, people holding those sorts of jobs in dictatorships are not natural heroes to the democracies. Indeed, they are people for whom our sort of society feels little sympathy and quite a lot of prejudice: that is surely true of the former German officer-class as a whole. Such people therefore tend to be written about unsympathetically or dismissed.

The second factor is harder to define, because it is a void. I mean the gulf of understanding that divides those who have lived in a modern dictatorship and those who have not: it is like the gulf in experience between those who have and have not suffered a bad illness. Only a few gifted writers, like Koestler and Orwell, have managed to make this imaginative transposition. The rest of us just cannot put ourselves in the position of citizens of the Third Reich or of the USSR—whether we realize our limitation or not.

In mid war, June 1942, a report was delivered to the Foreign Secretary in London by an eminent Dutch churchman: it came from the German Opposition and had been brought out by Trott. Eden, the Foreign Secretary, wrote a note to his Cabinet colleague, Cripps, who had enquired about it. He said he did not intend to respond to 'these people'. 'Our view [he wrote] is that until they come out into the open and give some visible sign of their intention to assist in the overthrow of the Nazi régime, they can be of little use to us or to Germany.'[8] Eden was not expressing a merely personal opinion. This view of the German Opposition was not opposed by the Foreign Office or by the experts on Germany engaged in directing our propaganda or by anyone else in Government, as far as I know, except by Cripps. Yet, what could the words 'come out into the open' to show an 'intention to assist in the

[8] FO/C. 5428/416/G, PRO, 18 June 1942.

overthrow' of the regime possibly mean, in terms of Nazi Germany? They apply to open societies. In the Third Reich, such terms meant immediate arrest and execution. And how much 'assistance' could any anti-Nazi resisters ever have given a British government that would not answer their messages or give them the kind of help that they gave to every other resistance movement in Europe?

Eden goes on to write that Trott and his friends 'have never been quite able to pay the price of their convictions and resign from the service of the Nazi régime'. Did he mean that they were too career-minded and cautious to give up their jobs—although willing to carry treasonable documents to enemy embassies in wartime and eventually be hung for it?

It may well be genuinely impossible to understand the realities of situations such as theirs, just as we may not be able to imagine the milder realities of political life inside the Soviet Union and in other authoritarian states today. However, our chance of crossing these gulfs would be increased if we were aware of our ignorance. It is also necessary to regard the people on the other side of them as being, perhaps, our moral equals. The patronizing tone of Eden's letter is its most ludicrous feature.

3

WIDERSTAND—RÉSISTANCE: THE PLACE OF THE GERMAN RESISTANCE IN THE EUROPEAN RESISTANCE AGAINST NATIONAL SOCIALISM

KLEMENS VON KLEMPERER

THERE is something encouraging about the invitation to talk about resistance. The challenge of the Third Reich, the subject of this series, was a momentous one. National Socialism was far more than a German monstrosity, far more than the disastrous resolution of the crisis of the Weimar Republic. Its sway was not restricted to Germany alone. Nazism in Germany was but the most visible symptom of a general European malady. Already long before the German troops swept across Europe it had its hold over a demoralized Continent which was plagued by political fragmentation, economic depression, and self-doubt, and all too ready to appease Adolf Hitler if not actually to embrace a new order. Nazi dominion and occupation, initially at least, were given a warmer welcome than historians have hitherto recognized. They brought about confusion, treachery, surrender, but only as a last resort resistance. The cruellest tyranny, we have learned in the course of our century, is the one which comes in the guise of light, charity, or historical necessity; it constitutes, as Pastor Dietrich Bonhoeffer wrote in a Nazi jail, the 'great masquerade of evil'.[1] He was referring, of course, to the Nazi tyranny, which thus played havoc not only with the Germans but with almost all the peoples of a troubled Continent.

It is all the more encouraging, then, that there were men and women who knew their minds and were determined to 'stand fast'—this is again a term which Bonhoeffer used[2]—to say 'No', and who engaged the oppressor in an unequal struggle. In those dreadful years from 1933 to 1945 they stood for courage, for love of country, and ultimately for Europe and for the affirmation of the dignity of man. Dignity, I hasten

[1] Dietrich Bonhoeffer, *Letters and Papers from Prison* (enlarged edn., New York, 1972), 4.

[2] Ibid.

to add, was a matter of not yielding to the lure of the new order, of preventing subjection to and destruction by it and ultimately, of course, of putting up resistance to it.

The men and women who made the decision to follow their consciences, whether as solitary witnesses or in concerted action, obviously did not win the war; the war was won by the armies of the Grand Alliance. But without these men and women and without resistance the armies would have had nothing worth liberating. These people kept, as the title of a book on the German Resistance suggested, the 'flame of freedom' burning.[3] Resistance as it materialized across the European Continent in the course of the war, its deeds, its thoughts, and its plans for a European order constituted the ultimate and most positive challenge to the Third Reich.

My particular topic is the German Resistance—that is, a re-examination of the German Resistance. Might I remark at this point that I find it singularly moving and appropriate that it is in Oxford that the re-examination is taking place. In Oxford, Adam von Trott, to whosc memory these lectures are dedicated, virtually found a second home; in Oxford, moreover, Adam von Trott first met Helmuth von Moltke, with whom he was to co-operate closely in the Kreisau Circle, one of the centres of the German Resistance.[4] Indeed it was with Britain that Trott, like his fellow conspirators, maintained contacts before and throughout the war against increasingly insuperable odds. And it is in Britain that he remains to this day in a strange—might I say undeserved—twilight, admired by some and distrusted by others. A better understanding of the German Resistance, whatever its shortcomings, should lead to a better understanding of the predicament in which Adam von Trott found himself as a pioneer in that hazardous and virtually unprecedented business of Resistance foreign policy. The Trott problem, I suggest, is essentially the problem of the German Resistance, and vice versa. Let me state the issue candidly and vigorously: it is the problem, first, of the underestimation and misunderstanding with which the Allied countries received the persistent fcclcrs extended by the German Resistance before and during the war; it is the problem, also, of the subsequent historical treatment of these efforts.

[3] Eberhard Zeller, *Geist der Freiheit, der zwanzigste Juli* (Munich, 1963), tr. R. P. Heller and D. R. Masters, *The Flame of Freedom, The German Struggle against Hitler* (London, 1967).

[4] A. L. Rowse, *All Souls and Appeasement. A Contribution to Contemporary History* (London and New York, 1961), 95.

Certainly the distinction between the European Resistance movements and the *Widerstand*, that is the German Resistance against Hitler, became a commonplace both in the politics of the Second World War and in the historiography, and it survives even today. The former, for which the French word *Résistance* has become a general designation, have been widely seen as engaged in a common struggle to free their countries from occupation and to reinstate some form of national integrity and human rights. The 'common struggle' against the enemy was the term used in the summer of 1940 by Hugh Dalton, the British Minister of Economic Warfare, in his now famous Secret Memorandum to the War Cabinet entitled 'The Fourth Arm' outlining the need for subversion on the Continent and for the creation of a British agency, the later SOE (Special Operations Executive), which, in co-operation with the army, navy, and air force was devised to give support, psychological, political, material, to all existing and future resistance movements.[5] It might be interjected right here that at that particular moment certainly no European resistance movement did exist; only the German Resistance did. And while in the early phases of the war London became 'the arsenal, the banker and the headquarters'[6] of the European Resistance, in the later phases the United States and Russia stepped in with their resources.

It has generally been assumed that the underground movements in the occupied countries were widespread. This assumption was voiced by Winston Churchill in a New Year's message of 1945 to the Danish Resistance in which he said, 'when we in Britain speak of the *Grande Alliance*, we mean not only the armies, navies, and air forces of the United Nations, we mean also the resistance movement throughout Europe, whose members have played so gallant a part in this total war against a brutal and unscrupulous enemy'[7]—a fitting tribute to a noble venture. This assumption certainly fed the mystique which dominated the thinking about the European Resistance during the years immediately following the war—a mystique of valiant underground armies, in occupied countries, backed up by massive popular support.

In all this the German Resistance has found no place. When, not too long ago, the *Times Literary Supplement* referred to the German

[5] Cf. Hugh Dalton, *The Fateful Years. Memoirs 1939–1945* (London, 1957), 366 ff.; Jørgen Hæstrup, *European Resistance Movements, 1939 –1945: A Complete History* (Westport, Conn., and London, 1981), 10 f., 38 f., 348 ff.

[6] Henri Michel, *The Shadow War. European Resistance 1939–1945* (New York, 1972), 53.

[7] Børge Outze (ed.), *Denmark during the German Occupation* (Copenhagen, 1946), 84.

Resistance as 'one of the non-events of the Twentieth Century',[8] it was in a macabre sense not so far off the mark. Certainly it was a non-event for the Allies, and met with close to no understanding and recognition. And if it was at all acknowledged to have existed, its motives and objectives were called into question. Neville Chamberlain likened Ewald von Kleist-Schmenzin, one of the German emissaries who came over to London in 1938 to warn against Hitler's plans of aggression and to urge Britain to take a strong stand, to the Jacobites at the court of France in King William's time.[9] The professionals of the Foreign Office dismissed such approaches as that of the young lieutenant colonel of the German General Staff Gerhard Graf von Schwerin in January 1939 for their 'gross treasonable disloyalty'.[10] Of course in the German case resistance did border on treason—treason, however, commmitted against a regime, the Nazi regime, which was violating the true interests of the German nation and indeed of humanity. Why then the scorn? And Colonel Hans Oster of the *Abwehr*, the German Army Intelligence, who was deeply involved in the conspiracy against Hitler, and who kept the Dutch military attaché informed about the dates of the German offensive in the West, was promptly and with equal scorn dismissed by Holland's Commander-in-Chief, General H. G. Winkelmann, as an 'erbärmlicher Kerl'.[11]

The very distinction between Germans and Nazis, fundamental to the self-image of the German Resistance, faded on the Allied side as the war went on, and the much heralded plot did not seem to materialize. Not surprisingly, perhaps, members of the German opposition who kept seeking a hearing from the Western Allies throughout the war met consistently with rebuff. No Allied help was extended to them; moreover they ran into the wall of 'absolute silence'[12] and 'unconditional surrender.'[13] And even after the failure of the coup of 20 July 1944

[8] 'Troubled Resistance', *The Times Literary Supplement*, 27 Mar. 1969, 322.

[9] *Documents on British Foreign Policy 1919–1938*, 3rd ser. ii. *1938* (London, 1949), 648.

[10] FO 371/22963/1291/15/18.

[11] Institut für Zeitgeschichte (Munich), ZS 1626, Sas, 16; cf. Romedio Reichsgraf von Thun-Hohenstein, *Hans Oster. Versuch einer Lebensbeschreibung* (Diss. Kiel, 1980), 11, 215.

[12] Foreign Office Secret 'Summary of Principal Peace Feelers September 1939–March 1941', 14 Apr. 1941, 2: 'Our attitude towards all enquiries and suggestions should henceforth be absolute silence'; FO 371/26542/C4216/610/G.

[13] The formula agreed upon by President Franklin Delano Roosevelt and Prime Minister Winston Churchill at the Casablanca Conference, 12–24 Jan. 1943, which remained the basic policy directive of the Allies throughout the war.

Winston Churchill's report to the House of Commons, curt and acid and disdainful, spoke of 'the highest personalities in the Reich . . . murdering one another', characterizing the plot as a manifestation of 'internal disease'.[14] A British Intelligence Report of December 1945 insisted that 'while one cannot but admire the plotters' . . . desire to get rid of Hitler, the motives for it and the programme which they hoped to put into practice after it, both are very far from harmonising with our ideas of a genuine movement of liberal resistance'.[15] The assumption in the Report, then, was that the men of the German Resistance, notwithstanding their heroism, acted without a valid mandate; that they fought tyranny, but not in the name of freedom; that they fought evil, but not in the name of universal values.

A similar ready indictment of the German Resistance appears in the work of historians in Anglo-Saxon countries. It is striking that in most of their general accounts of the European Resistance the German Resistance is either not included or at best given a marginal place: 'even in their German homeland . . . resisters were to be found'.[16] Besides it was, so the story goes, a 'generals' plot' and no more. The German Resistance, then, has gone down in history as at best a predominantly 'ethnocentric' matter, as a manifestation of that German *Sonderweg*, as historians call it nowadays,[17] the historical and political path which in the past two centuries made the Germans diverge from the common Western experience and which in many ways accounts for the uneasy relation of the Germans toward freedom and modernity. And if Hitler was a culmination of that development, so, the argument goes, was the Resistance against him. This assumption has provided grounds for the thesis that there was a basic ideological identity between the plotters and the Nazis; it had led to the hard question: how can one rebel against something with which one is identified? This question has been brutally, and I might say, absurdly, answered by an American critic: 'the rebels had no ideas, but died because they could not bear the stench of their own deeds'.[18]

[14] Winston Churchill, 2 Aug. 1944, *Parliamentary Debates, House of Commons*, 402 (1943/4), col. 1487.

[15] GSI, HQ, British Troops, Berlin, Intelligence Summary No. 24, 17 Dec. 1945.

[16] M. R. D. Foot, *Resistance. An Analysis of European Resistance to Nazism 1940–1945* (London, 1976), 4.

[17] Cf. for a discussion of this issue David Blackbourn and Geoff Eley, *Mythen deutscher Geschichtsschreibung. Die gescheiterte bürgerliche Revolution von 1848* (Frankfurt and Berlin, 1980) and the review of the book by Wolfgang J. Mommsen in *Bulletin* (German Historical Institute, London), autumn 1981, Issue 8, 19–26.

[18] Henry M. Pachter, 'The Legend of the 20th of July, 1944', *Social Research*, 29 (spring 1962), 113.

In the Federal Republic of Germany there has appeared an authoritative scholarly literature vindicating the *Widerstand*, but it has on the whole remained without popular resonance.[19] A Beck, a Goerdeler, a Moltke, a Trott, a Stauffenberg have found no place in the hearts or even minds of most Germans. Meanwhile there has appeared in recent years a revisionist school of writing which, in tune with British and American scholarship, has called into question the motives and objectives of the German conspirators who, far from being the voice of 'eternal Germany' and far from being propelled by lofty principles, never got beyond defending their respective professional preserves as they were encroached upon by Hitler. General Ludwig Beck, then, in the *Widerstand* looked up to as the 'sovereign', was, in fact, but the head of a class-inspired *fronde*.[20] It should be added that in West German resistance studies the focus has been until recently entirely too much on the conservative Resistance and on 20 July 1944, when the Resistance surfaced most visibly though of course unsuccessfully, and too little on the other manifestations of resistance, like the many people of humble station often who entirely on their own went for their convictions into a solitary death, the incidence of popular resistance (*Volksopposition*), and the resistance from the Left—especially the Communist Resistance, which in the Federal Republic has been relegated all too readily to the realm of espionage and treason.

We have, then, a mystique of the *Résistance* here and an anti-mystique of the *Widerstand* there. But the dust has now settled over wartime emotions, exigencies, and priorities, and our understanding of the problems of Resistance all over Europe has advanced so that we can re-examine this rigid juxtaposition of *Résistance* and *Widerstand*. The challenge of this hour should be to redress the balance between the two, to demythologize both.

I should now make four general points about the nature and condition of resistance against National Socialism which may at first be disquieting. They will help me to close the gap between *Résistance* and *Widerstand*.

[19] Cf. among the many works especially Hans Rothfels, *The German Opposition to Hitler. An Assessment* (London, 1970); Peter Hoffmann, *The History of the German Resistance 1933–1945* (Cambridge, Mass., 1970); Ger van Roon, *Widerstand im Dritten Reich. Ein Überblick* (Munich, 1979).

[20] Cf. especially Klaus-Jürgen Müller, *General Ludwig Beck. Studien und Dokumente zur politisch-militärischen Vorstellungswelt und Tätigkeit des Generalstabschefs des deutschen Heeres 1933–1938* (Boppard am Rhein, 1980).

(1) The Europeans did not, as the folklore would have it, rush into resistance immediately. Resistance everywhere came late. The trauma of invasion, defeat, and occupation outside Germany was in its way as disarming as was the seizure of power in Germany. Paralysis was the prelude to resistance all over Europe, not in Germany only, and, save perhaps in Holland, bewilderment, relief, despair were at first the order of the day, a range of emotions which fed into initial attitudes of *attentisme* and which preceded the taking of distinct political positions like resistance or collaboration.

(2) The gulf between resistance and collaboration was not as wide as the stereotypes prevalent immediately after the war might suggest. Certainly Marcel Ophuls's sobering film *Le Chagrin et la pitié*[21] has brought home that lesson, and recent work on the European Resistance has made us face up to it, extending the range of both resistance and collaboration so that they seem to converge upon each other. There is no divide between the more cautious or limited instances of resistance and the less outspoken and exposed cases of collaboration. How, for instance, should we assess a brave man like Dietrich Bonhoeffer's concern with protecting the baptized Jews rather than all embattled Jews; or the Ecumenical Movement's failure, in the early years of Nazi rule, to support unequivocally the Confessional Church in its struggle with Nazism? or the course which King Christian X of Denmark chose, to shield his countrymen by staying at home? or the efforts of the Netherlands Union in Holland to 'gather all patriots . . . in loyal attitudes toward the occupying power'[22] and thus to prevent the appointment of a Fascist puppet government headed by Anton Mussert? or the cunningly calculating policies of the *baillis* of the Channel Islands? or the careful strategies of the many German leaders, like the mysterious Admiral Canaris of the *Abwehr* or State Secretary von Weizsäcker in the Foreign Office, who stayed in office and by publicly performing their often iniquitous tasks, served as a front for all kinds of resistance feats by their subordinates? In turn some of the most daring resistance feats, especially in the German setting, had to be performed under 'official' auspices, like the many journeys abroad of Trott, Moltke, and Bonhoeffer, or the attempt of Kurt Gerstein, the 'spy of God', a deeply religious

[21] *The Sorrow and the Pity*. A film by Marcel Ophuls (New York, 1972). Cf. also especially Werner Rings, *Life with the Enemy. Collaboration and Resistance in Hitler's Europe 1939–1945* (New York, 1982).

[22] Werner Warmbrunn, *The Dutch under German Occupation 1940–1945* (Stanford, 1963), 133.

man, to expose Hitler's euthanasia programme by joining the SS and its extermination unit.[23] 'Almost everyone', wrote Louis de Jong, the historian of the Dutch Resistance, 'practiced resistance and collaboration at one and the same time.'[24]

(3) Resistance, once it materialized, was primarily a matter of individuals and not groups. 'Character, not . . . class origin', observed M. R. D. Foot quite correctly, 'made people into resisters'[25] and for that matter collaborators; not class, and he might have added 'not political grouping'. So far too much deference has been paid in the histories of the Resistance to the 'representational theory of resistance'[26] which insists upon correlating resistance with specific political, social, or religious groups. It certainly cannot tell the whole story. These groups tended not to go much beyond the point of what Richard Löwenthal has called 'social refusal'[27] inasmuch as they primarily sought to fend off intrusion into their respective realm or, as in the case of the French Communist Party, went from one extreme—deterrence of its members during the time of the Nazi–Soviet Pact from resistance activities—to the other, active resistance thereafter.

The road to resistance was in fact often an indirect and incidental one; all kinds of motives led to the decision to resist: loneliness, vanity, friendship, hankering after adventure, anger, pride—not always elevated motives, but always, sooner or later, the awareness of the evil that was to be fought. Resistance was above all a lonesome business. It meant a lonesome decision for Marc Bloch, who left his wife and six children and his work to join the *Résistance*; and Adam von Trott on his journey to the Far East in 1937–8 pondered over the question whether to join the conspiracy against Hitler. Also, all kinds of people, men and women, high and low, of all kinds of backgrounds and political persuasions from

[23] Pierre Joffroy, *Der Spion Gottes. Die Passion des Kurt Gerstein* (Stuttgart, 1972); cf. also Saul Friedländer, *Kurt Gerstein ou l'ambigüité du bien* (Paris, 1967).

[24] Louis de Jong, 'Zwischen Kollaboration und Résistance' in Andreas Hillgruber (ed.), *Probleme des Zweiten Weltkrieges* (Cologne, 1967), 252.

[25] Foot, *Resistance*, 11.

[26] Peter Hüttenberger, 'Vorüberlegungen zum "Widerstandsbegriff" ' in Jürgen Kocka (ed.), *Theorien in der Praxis des Historikers* (Göttingen, 1977), 118.

[27] 'Gesellschaftliche Verweigerung': Richard Löwenthal, 'Widerstand im trotalen Staat' in Richard Löwenthal and Patrick von zur Mühlen (eds.), *Widerstand und Verweigerung in Deutschland 1933–1945* (Berlin and Bonn, 1982), 14, 18 ff. A distinction similar to the one made by Löwenthal between *Widerstand* and *gesellschaftliche Verweigerung* had been made earlier by Martin Broszat between *Widerstand* and *Resistenz*; but especially in a comparative treatment of problems of resistance this distinction is linguistically misleading; cf. Martin Broszat, 'Resistenz und Widerstand' in id. *et al.*, *Bayern in der NS-Zeit*, iv (Munich and Vienna, 1981), 691–709.

the far right to the far left came together in the Resistance. In France the *Résistance* embraced a Colonel de la Rocque, the onetime leader of the Croix-de-Feu, as well as a Jacques Duclos, the leader of the clandestine Communist Party. In Germany the spectrum was no less broad, stretching from someone like the repentant Nazi Berlin Police President Wolf Graf Helldorf, a drunken vainglorious lout of sorts, to the saintly Dietrich Bonhoeffer, and from the right all the way to the far left. For all resistance ultimately meant a leap into the unknown, into an unprecedented experience of extremity that calls for what Dietrich Bonhoeffer called 'responsible action' and readiness for suffering. Might I in this connection quote from the memoir of one of Trott's oldest friends, who wrote 'resistance means continually to come to terms with one's conscience and to make daily the decision which determines action and keeps the aim in sight, even if always the decision must be made and the way must be gone alone'?[28] This is well said, and speaks of an intimate understanding of what motivated Trott and his fellow conspirators.

(4) Resistance against National Socialism was everywhere and at all times a matter of the few.[29] The old 'official version', according to which Resistance movements in the occupied countries were patriotic majorities opposing the invader, no longer holds. It is now understood that certainly before the Allied invasion of June 1944, the numbers of, say, Frenchmen and Dutchmen actively engaged in resistance were minuscule, as were those of the Germans in the *Widerstand* all along. But in this context figures are cruel and misleading. Brave men and just men are rare in all situations, but are especially so in situations where the oppressor has practically total control.

In its own way, then, the Resistance in Germany against National Socialism shared with the Resistance movements in the occupied countries some important characteristics: its lateness—or, as Pater Max Pribilla called it, the 'weakness'—in beginning;[30] the often close correlation with compliance, the German equivalent of collaboration; the wayward approaches to resistance; the isolation and loneliness of a minority. These similarities are more than incidental; they are determined by the nature of the oppression. In the landscape of totalitarianism, we now understand, the possibilities and forms of resistance

[28] Ingrid Warburg-Spinelli, 'Die Reise nach Deutschland—Juli/August 1982', typescript, 1982.

[29] Rings, *Life with the Enemy*, 211 f.

[30] *Vollmacht des Gewissens*, i (Europäische Publikation e.V. (ed.), Frankfurt and Berlin, 1960), 21 f.

remain circumscribed everywhere, and they were so in the occupied countries as well as in Germany. The German case, then, I suggest, was a variant on the general European one rather than a deviation.

It is worth recalling, however, the often overlooked fact that dictatorship in Germany, being, as Richard Löwenthal put it, 'native',[31] came closer than elsewhere to the model of total control; control through terror backed up by mass manipulation. The consensual quality of Nazi control was one of its most vexing paradoxes. It made oppression all the more humiliating and—what matters in this context—it rendered the task of resisting all the more difficult. Resistance in Germany had to be staged without that 'social support'[32] which was available at least latently to the *Résistance*. This distinction is important. Far from inviting us in any way to minimize the hardships of the *Résistance*, it may serve to explain many, though not all, of the special difficulties which the German Resistance faced on its conspiratorial path—a *caveat* to its detractors.

It is to these difficulties that I now propose to turn. At this point we should remember that in Germany not even the designation *Widerstand* was current among those who resisted Hitler. One of the more moving interviews I have conducted with the few survivors was with one of the most fearless members of the plot to free Germany of the tyrant, a great lady. She said to me insistently: 'Don't talk about *Widerstand*. We did not think of ourselves as being part of a *Widerstand*. We merely sought somehow to survive in dignity.'[33] If there was no word for it, did the thing itself exist? My assignment here being not to philosophize but to bear witness and to analyse, I must give names to events that I identify in the past. Thus, with all due respect to the lady I so much admire, we must assume that there was a German Resistance, that there was a *Widerstand* against Hitler. I might suggest, however, that the absence of a common nomenclature, as also the absence of a Resistance movement, in Germany may have made the decision to resist, to act, all the more a matter, not of routine and convention, but of a personal decision and of self-discovery.

It would be foolish, of course, to ignore the peculiarities of the German case. The diminishing awareness in modern Germany of the

[31] Löwenthal, 'Widerstand', 12.

[32] For the importance of 'social support' in oppressive situations see Barrington Moore Jr., *Injustice. The Social Bases of Obedience and Revolt* (London, 1978), 97. Foot writes about the importance of the 'feeling of participation'; Foot, *Resistance*, 5.

[33] Interview with Marie-Luise Sarre, 8 Mar. 1978.

workings of natural law and the concentration on the prerogatives of the *Obrigkeitsstaat* furnished less than ideal preconditions for resistance. In fact Dietrich Bonhoeffer, writing from a Nazi jail, noted that the Germans in their long history had 'come to face up to the necessity and power of obedience', adding that as a result they had been wanting in what they call 'civil courage'.[34] *Civilcourage* was not exactly a German national trait.

Neither can it be ignored that the very segments upon which, after 1933, the primary burden of resistance fell had either failed to stand up for the Weimar Republic, like the Social Democrats, who already before Hitler's seizure of power had shown that they were unable or unwilling to resist a threat of right-wing dictatorship, or had actively helped to undermine the Republic, like the Communists, who punctuated their fight against fascism by their fierce denunciation of the Socialists as 'social fascists' and even made common cause with the National Socialists. Nor should it be overlooked that after the seizure of power the masses became rapidly Nazified. Under these circumstances individual expressions of defiance, personal acts of quiet heroism, simply could not feed a broader current of popular resistance. The Social Democratic organization was too suspect and too exposed to conduct organized resistance within Germany. The only practical alternative left to the Party leadership was emigration. The party leaders who stayed on—men like Wilhelm Leuschner and Julius Leber—came to understand that underground activity without co-operation by the army was doomed to failure.[35] Meanwhile the Communists, who initially thought of staging popular resistance on a broad basis, became an easy target for Nazi terror. It was only after they had suffered staggering losses among their leadership that the Communists regrouped and shifted to resistance by means of illegal activity by small cells,[36] thus laying the foundations for continued proletarian underground activities even though they could expect little more than local successes and the sheer survival of their organization.

The indifference, if not hostility, to the Weimar Republic of the middle and upper layers of society and in particular of the old establishments, the army, the churches, and the public service, is all too well

[34] Bonhoeffer, *Letters*, 5 f.

[35] Cf. on this subject Hans-Joachim Reichardt, 'Möglichkeiten und Grenzen des Widerstandes der Arbeiterbewegung', in Walter Schmitthenner and Hans Buchheim, *Der deutsche Widerstand gegen Hitler* (Cologne and Berlin, 1966), 169–213.

[36] Cf. Hermann Weber, 'Die KPD in der Illegalität', in Löwenthal and von zur Mühlen, *Widerstand*, 83–101.

known, as are their differences with National Socialism. It stands to reason that, considering the conditions in Nazi Germany, these groups, if any, would emerge as nuclei of opposition.[37] Circumstances were such that the only potentially effective resistance in Germany was 'resistance without the people',[38] as Hans Mommsen has called it. In Germany in particular, then, the convergence of resistance and compliance, or 'opposition in the service', as State Secretary Ernst von Weizsäcker called it, has been almost imperative. In many instances the continued service to the Nazi State seemed justified by the calculation that it might be effective in causing obstruction and averting the worst excesses of the regime. But in most cases staying in office, that is practising 'feigned co-operation',[39] was a necessary cover for conspiracy. Most of those who resisted did so, indeed had to do so, from within the old establishments and thus had to take upon themselves the burden of split existence which is the law of conspiracy.

Those in public office had, of course, the choice of staying on in office or resigning. Carl Goerdeler resigned as Lord Mayor of Leipzig in protest against the anti-Semitism and the Church policies of the Nazis, and General Beck resigned over Hitler's plans of aggressive warfare. Weizsäcker and Canaris, who chose the path of the many who stayed on, could thus, it can be maintained, safeguard their conspiratorial effectiveness better than the others. It was in face Beck himself who urged Weizsäcker to stay on.[40] But who is to judge which of the two options was the braver and more honourable one? Resignation, it could be argued, was tantamount to abandoning ship, whereas holding on to the office meant perseverance; but with equal validity resignation could be interpreted as an exemplary gesture—and if, as Beck had expected, the other army commanders had followed suit, the gesture might have been effective too—and staying on as a symptom of timidity. Both courses have been judged with excessive intensity and all too frequently without regard for the unique dilemmas which the German opposition saw itself confronting. No facile moralizing can do justice to the problems of resistance against a totalitarian regime like the Nazi one. In

[37] Cf. on this subject Hans Mommsen, 'Gesellschaftsbild und Verfassungspläne des deutschen Widerstandes' in Schmitthenner and Buchhéim, *Der deutsche Widerstand*, *Vierteljahrshefte für Zeitgeschichte*, 3 (1955), 297–310.

[38] Mommsen, 'Gesellschaftsbild', 76.

[39] Warren E. Magee, 'Opening Statement for Defendant von Weizsäcker', 3 June 1948, *Trials of War Criminals before the Nürnberg Military Tribunals*, *vol.* xii (Nuremberg, Oct. 1946–Apr. 1949), 241.

[40] Ernst von Weizsäcker, *Erinnerungen* (Munich, 1950), 173.

fact both courses, resigning and staying in office, were often the outcome of agonizing self-scrutiny and both were potentially honourable.

But there were in Germany all too many cases of compliance with evil and all too few cases of unqualified opposition. The incidence of timidity is undeniable, especially on the part of the generals, who, if any group, were in a position to act decisively. But they saw themselves bound by their 'soldier's oath' and were disdainful of lending their hands to so lowly a business as rebellion. Similarly striking is the meekness of ever so many princes of the churches, who supposedly were the guardians of Christian ethics in the face of an obvious evil. As institutions, the army, the churches, the public service offered no resistance. At best they exercised 'social refusal' in order to protect the residues of social pluralism against the claims of total dominion. Thus they each acted separately, not in conjunction with one another, and primarily to protect their own prerogatives. They did not as institutions face up to the political and moral challenge of Nazism.

Genuine resistance is of course more than a matter of 'degrees of non-conformity'.[41] In Germany as elsewhere it was a matter above all of personal decision to stand fast and to fight the evil, a matter of 'self-discovery' and 'self-renewal', as Richard Cobb has rightly observed. It was a 'personal state of war'[42] which eventually reconnected many of the individuals with new-old groupings like the Communists, rejuvenated after their virtual decimation, or various left-wing splinter groups like the one called 'Neubeginnen', which actually antedated the seizure of power and which developed intense resistance activities; it was this group, incidentally, with which Trott had close ties. Others were brought together in new formations, in Germany especially in the form of circles of friends like the Kreisau Circle around Moltke and youth groups like the Edelweiss Pirates and the so-called 'White Rose' group of students and their teachers in Munich. But in Germany, unlike in the occupied countries, it never came to the formation of a unified resistance movement, and the burden of resistance fell all the more on individuals or small groups.

The most vexing problem in Germany was, of course, that of the legitimacy of resistance. The resistance movements in France, in Holland, in Norway were directed against occupation and oppression by

[41] Cf. Rothfels, *The German Opposition*, 27ff.

[42] Richard Cobb, 'A Personal State of War', *The Times Literary Supplement*, 10 Mar. 1978, 270.

a foreign power, and their struggle for liberation was a clear and unquestioned assertion of national interests as well as of human rights. But the German Resistance against Hitler had no such clear mandate. The plot was out of step with the immediate national effort, indeed with the waging of the war, and was subversive of what appeared at least to be the national interest. For the success of the plot might have meant defeat for the fatherland. The plotters, then, were engaged in an enterprise in which heroism bordered on treason and resistance on defeatism. Rebuffed abroad and wholly isolated at home, they persisted in seeing themselves as patriots even while they risked the defeat of their own country.

The isolation of the Germans was ever so urgently conveyed by Helmuth von Moltke in a letter of March 1943 which he managed to send via a Swedish ecumenical connection to his old friend Lionel Curtis in England: 'People outside Germany do not realise the following handicaps under which we labour and which distinguish the position in Germany from that of any other of the occupied countries: lack of unity, lack of men, lack of communications'; and he added: 'In the other countries suppressed by Hitler's tyranny even the ordinary criminal has a chance of being classified as a martyr. With us it is different: even the martyr is certain to be classed as an ordinary criminal. That makes death useless . . . '.[43]

But it did not after all make death useless. It certainly spurred the Resistance groups on to break out of their isolation by intrepid efforts, despite the Allied policies of 'absolute silence' and 'unconditional surrender', to reach the outside world.

We see, then, that the particular circumstances under which the *Widerstand* laboured at home, especially the regime's totality of control and mass support, the ambiguity of its own mandate, and its isolation, account largely for its divergence from the resistance movements in the occupied countries. In the German case of course the 'weakness of the beginning' was fatal. Allowing for all the complexity and ambiguity of the *Widerstand*'s mandate, which delayed and inhibited swift and concerted action, it should have been clear by 1934, that is by the time of the Blood Purge of 30 June, that the time for resistance had come. In moral and political terms the hands of the German Resistance would have been infinitely stronger in 1934 than in July 1944, the time of Stauffenberg's unsuccessful coup. Certainly the conspiratorial course of

[43] Michael Balfour and Julian Frisby, *Helmuth von Moltke. A Leader against Hitler* (London, 1972), 216, 220.

the Widerstand was an appropriate response to totalitarian pressure, and it dictated the 'feigned co-operation' with the establishment which remained so much of a puzzle to the outside world.

Now, the predominantly conservative stance of the German Resistance calls for re-evaluation. The misgivings about it, so frequently voiced, should not be lightly dismissed. The fact is that the Resistance that took shape after the organized workers' movement had been broken was predominantly aristocratic and conservative. Its outlook, especially that of the so-called 'Seniors',[44] or group around Beck and Goerdeler, was essentially restorative. Moreover their circle included men who maintained close ties with the SS in the dubious expectation of playing Himmler off against Hitler, thus using Satan to drive out Beelzebub.

On the other hand the so-called 'Juniors' of the *Widerstand*, and in particular the members of the Kreisau Circle, were distinctly in search of a new beginning. The Kreisau Circle's very composition gives us a good idea of the generously explorative and humanitarian spirit that animated it. It included noblemen and commoners, Protestants and Catholics, conservatives and Socialist labour leaders. Uncommitted to party creeds and rejecting one-sided doctrinaire positions, these men sought to learn from each other and to prepare the ground for a new synthesis which would heal all the divisions that had plagued Germany in the past. They firmly dismissed as reactionary Goerdeler's views on domestic affairs and as nationalistic his views on foreign affairs. They groped for a blending of conservative and socialist ideas.

But it is striking that even the Socialist members of the Kreisau Circle, like Wilhelm Leuschner and Julius Leber, not only dissociated themselves from the 'outworn' party structure of the Weimar Republic but also rejected egalitarian mass democracy.[45] Indeed Stauffenberg and his brother went one step further when, in their draft for the oath which was to be sworn after the *coup d'État*, they put into it a reference to 'the lie', as they called it, 'that all men are equal'.[46] In fact, the conspirators

[44] The distinction between the 'Seniors' and 'Juniors' among the resisters appeared first in Ulrich von Hassell's diaries: Ulrich von Hassell, *Vom anderen Deutschland*, (Frankfurt, 1964), 260. The two groups met on 8 Jan. 1943 in Berlin with the 'Seniors' being represented by Beck, Goerdeler, Hassell, Johannes Popitz, Jens Peter Jessen, and the 'Juniors' by Moltke, Peter Graf Yorck von Wartenburg, Fritz-Dietloff Graf von der Schulenburg, Eugen Gerstenmaier; cf. also Hoffmann, *History*, 360.

[45] Cf. especially Annedore Leber, *Das Gewissen steht auf* (Berlin and Frankfurt, 1956), 98 and Julius Leber, *Ein Mann geht seinen Weg, Schriften, Reden und Briefe* (Berlin and Frankfurt, 1952), 220 ff.

[46] Joachim Kramarz, *Stauffenberg. The architect of the Famous July 20th Conspiracy to Assassinate Hitler* (New York, 1967), 185.

looked on National Socialism as an extreme instance of mass rule like Bolshevism. They were trying to devise a 'German way'[47] that should lie between liberalism and egalitarianism, between capitalism and collectivism—a way that should have brought about a renewed emphasis on Christian values, and more than that, a reintegration of man into his natural environment and neighbourhood.

The British Intelligence Report that I have cited earlier was, then, not that far off the mark with its reservations about the German conspiracy. Indeed critics, like Ralf Dahrendorf not too long ago, have seconded these reservations by arguing that in its distrust of plurality and conflict between political and social interests the German Resistance did not escape the pitfalls of social romanticism. This may well have been the case. But the conspirators' conservative tendencies may have been prompted, as has recently been suggested,[48] by the *Widerstand*'s need to plan for an uncertain post-war situation in which it figured a precipitate introduction of outright democratic reforms might bring about nothing short of chaos. A more traditionalist regime might therefore serve as a sensible transitional provision.

On balance the conservatism of the German opposition should not make us conclude that it was wanting in its advocacy of freedom and universal values. In its emphasis on federative polity rather than political parties and on solidarity rather than conflict, it moved well in the mainstream of the European conservative tradition and indeed in the tradition of Christian social thought. Its 'German way', in short, was by no means separatist; it was a 'conservative way' and a 'Christian way' as well. Moreover, it was in tune with the thinking prevailing all over Europe among Resistance groups of most shadings, that the return to the old pattern of party rule was out of the question and that liberalism and capitalism were not the proper prescription for the reconstruction of Europe.[49]

Therefore on both sides of the trenches, so to speak, the challenge of the Third Reich led to a rethinking of fundamental questions of state and society. On both sides there prevailed an awareness of standing on the threshold of a new age. All Resistance groups, whether on the right or the left, had a vision of the future. And if that of the *Widerstand* was, on

[47] Mommsen, 'Gesellschaftsbild', 161 ff.

[48] Peter Hoffmann, *Widerstand gegen Hitler. Probleme des Umsturzes* (Munich, 1979), 21, 60.

[49] Cf. Walter Lipgens, *Europa-Föderationspläne der Widerstandsbewegungen 1940–1945* (Munich, 1968), *passim*.

balance, distinctly more conservative—and I say 'conservative', not 'reactionary'—in nature, it was no less informed than the Resistance of the other side by the ethos of liberty and human decency. Indeed, embattled as the Germans were, they came to translate the Nazi challenge into a new understanding of the Christian message. Almost all conspirators were religious men. 'I have come to the conclusion,' wrote Trott to one of his English friends as early as the summer of 1936, 'that only a material renaissance of Christian law and ethics . . . can stem this tide which threatens to devour all we care for.'[50] Thus he spoke for the older generation as well as for his own, for the Protestants and Catholics, for the aristocrats and bourgeois, as well as also Socialists among his friends.

The foreign relations of the German Resistance, in which Trott played such an active part, also merit a second look. For it was in this area that the German Resistance saw itself firmly and consistently rebuffed and that the isolation of the *Widerstand* became particularly evident. The Allied Resistance movements needed no 'foreign policy'. They had their governments-in-exile as legitimate intermediaries and the direct support of the Allied intelligence services. And even though the relationships between them were not always easy, they were acknowledged relationships between allies. The German resisters were no allies. For them, caught in the 'vast jail', as a Socialist member of the opposition characterized the Nazi state,[51] the need to break out became all the more pressing. Doggedly, in the face of insuperable odds, a stream of German emissaries kept going abroad, to London especially and to Washington, and after the outbreak of the war to neutral capitals and to the capitals of occupied countries, in order to explore, to warn, to inform at the risk of committing treason, to press for a definition of peace terms, to negotiate with the 'enemy' conditions for facilitating a coup and territorial terms to be offered to a post-Nazi Germany, and last but not least to say to the world: 'We are here'.

Of course the German emissaries differed in their motivation, emphasis, and outlook. No doubt members of the older generation, men like Goerdeler and Ulrich von Hassell, were closely tied to traditional notions of German hegemonic thinking, thus, so it seemed, offering to

[50] Adam von Trott to Diana Hopkinson, summer 1936 in Diana (Hopkinson), 'Aus Adams Briefen', 95, copy in Leo Baeck Institute, New York.

[51] Leuschner: 'Wir sind gefangen in einem großen Zuchthaus'; quoted in Zeller, *Geist der Freiheit*, 95.

the foreign contacts, 'not another policy, but another partner'.[52] Even allowing for the fact that, say, Goerdeler's often shockingly nationalistic terms in part at least represented a way of saving the face of the opposition toward the generals and to secure public support after the coup, it is understandable that he should have been viewed in the British Foreign office as a 'stalking-horse for German *military* expansion'.[53] 'Are the stories which reach us of dissident groups in Germany genuine', a typical British Foreign Office document asked, 'or inspired by the German Secret Service?'[54] The truth of the matter is that in many instances they were both, since the German (army) Secret Service was deeply involved in the conspiracy against Hitler. Besides, the British were still haunted by the Venlo incident of 1939, when Nazi intelligence agents posing as German Resistance officers spirited two British Secret agents across the Dutch frontier into Germany. No doubt the Allied agencies, the British Foreign Office and the State Department, were ill prepared to deal with the unorthodox approaches of the German Resistance, and certainly the official Allied policy of 'unconditional surrender', defined at Casablanca in January 1943, locked the doors against any further dealings of the kind.

Nevertheless, Willem A. Visser 't Hooft, Secretary General of the World Council of Churches, who from his post in Geneva was deeply involved in the cross-currents of resistance networks, especially the Dutch and German ones, wrote in retrospect that the 'purely negative' Allied position was not justified. He argued that 'imaginative statesmanship' could have found a way of encouraging the German opposition, assuring it, without giving up the principles underlying Allied policy, that in planning for a future Europe there was a place for a Germany that had broken with Hitler.[55] Actually, Visser 't Hooft had close ties above all with Trott and with Bonhoeffer, that is with the younger generation of the German resisters who had freed themselves from hegemonic assumptions in favour of a broader European outlook. Indeed Trott's vision of a European federation based on self-determination and social justice, which he elaborated in November 1943 in response to a programme drawn up by John Foster Dulles for the

[52] Hermann Graml, 'Die außenpolitischen Vorstellungen des deutschen Widerstandes' in Schmitthenner and Buchheim, op. cit. 25.

[53] Cf. Sidney Aster, 'Carl Goerdeler and the Foreign Office' in A. P. Young, *The 'X' Documents* (London, 1974), 234.

[54] FO 371/39087/C 8865.

[55] W. A. Visser 't Hooft, 'The View from Geneva', *Encounter*, 23 (Sept. 1969), 94.

Federal Council of Churches,[56] coincided with the basic assumptions of a wide sector of the European Resistance movements. So did the Kreisau Circle memoranda[57] and even Goerdeler's 'eternal peace pact' of the late summer of 1943 in which neither Germany nor any other power was to be dominant.[58] Resisters in both Germany and occupied countries, then, were in agreement about the obsolescence of the old political forms, especially the sovereign state, and envisaged some sort of European federation.[59] Moreover, Trott's special advocacy of decolonization, regulation of minority problems, and disarmament, must be considered as radically innovative, if anything, compared with the programmes of most other European Resistance groups.

In fact, in the second half of the war the German Resistance sought direct contacts with Resistance movements in the occupied countries, especially the Dutch and Norwegian. This is what Moltke meant when in a letter of November 1942 he alluded cryptically to the importance of making headway with the question of 'the translation to the European level'.[60] He and his friends, especially Trott, indeed managed to establish a basic mutual trust with the movements in other countries and moreover some rudimentary co-ordination of efforts.[61] And in the spring of 1944 a meeting of representatives of the European Resistance movements took place in Geneva under the aegis of Visser 't Hooft at which two women of the German Socialist Resistance participated and which, furthermore, aimed at a federation in which Germany was to have full membership.[62] What the Foreign Office and the State Department had denied the German Resistance, the European Resistance movements were about to grant: trust and recognition. You may have noticed that I have just referred to the 'European Resistance

[56] 'Bemerkungen zum Friedensprogramm der amerikanischen Kirchen (Nov. 1943)', *Vierteljahrshefte für Zeitgeschichte*, 12 (1964), 318 ff.; cf. also Armin Boyens, *Kirchenkampf und Ökumene 1939–1945* (Munich, 1973), 222 f.

[57] Cf. van Roon, *Neuordnung im Widerstand* (Munich, 1967), 572 ff.

[58] Wilhelm Ritter von Schramm, *Beck und Goerdeler. Gemeinschaftsdokumente für den Frieden* (Munich, 1965), 255.

[59] Lipgens, *Europa, passim*; id., 'European Federation in the Political Thought of Resistance Movements during World War II', *Central European History*, (Mar. 1968), 5–19; W. A. Visser 't Hooft, *Memoirs* (London and Philadelphia, 1973), 177–81.

[60] 'Übersetzung auf das europäische Niveau', letter of Helmuth von Moltke to Freya von Moltke, 17 Nov. 1942.

[61] Cf. van Roon, *Neuordnung*, 323 ff.; letters of Dr C. L. Patijn, 18 June 1979, and Dr J. H. van Roijen, 12 May 1978, to the author; Theodor Steltzer, *Sechzig Jahre Zeitgenosse* (Munich, 1966); Arvid Brodersen, *Mellom frontene* (1979); interview of the author with Professor Arvid Brodersen, 22 Nov. 1980.

[62] Lipgens, *Europa*, 388 ff.; Visser 't Hooft, *Memoirs*, 177 ff.

movements'. At this point, with a shared vision and the direct dealings between *Résistance* and *Widerstand*, the distinction between the two has virtually disappeared and they have become one.

Therefore, the Allies' rejection of the German overtures, it must be argued, was up to a certain point only a matter of forestalling a suspected revival of German hegemonic aspirations, of distrusting the very unorthodoxy of all the feelers, and of adhering to the terms of the Grand Alliance with Russia which prohibited separate negotiations with the other side by any of the Powers. It was also, I should like to emphasize, a manifestation of the increasing rift between the Allied governments and the European Resistance movements at large. It has been pointed out by Walter Lipgens that, while early in the war the post-war planning of Great Britain and the US was geared towards one form or another of regional units within a broader European or world federation, thus conforming with the thinking and plans of the *Résistance*, in the latter half of 1943, the US took the lead in shifting its position and dropped these plans to accommodate Soviet demands for territorial concessions, which finally were agreed upon in Tehran in November 1943.[63] In the thinking of the Allied governments, annexation took the place of European order, annexation which, I hasten to add, served primarily the interests of the non-European superpowers, the US and Russia. The European Resistance, by contrast, remained committed to the future of Europe. A Dutch Resistance publication, for example, kept reiterating, perhaps too one-sidedly and defiantly, the position that in the future Germany must somehow be reabsorbed into the European community, adding: 'It is better to continue fighting for our ideal of a renewed Europe than to slip between the paws of the great beasts of prey in order to tear off our own part of the German cadaver.'[64] Once again, in the Resistance on both sides of the trenches, the challenge of the Third Reich elicited responses concerning European organization that were marked more by their affinities than by their differences. Once again, the rigid *Résistance–Widerstand* distinction must be dropped. At the cost of stating the obvious, let me say that both were historical events, and add that they were related ones; both were finally thwarted by the policies of the Great Powers. And perhaps in the future historians will take notice.

My remaining remarks, however brief, should be more than a post-script; they should certainly not be macabre although they have to do with the failure of both *Résistance* and *Widerstand*. About the former it

[63] Lipgens, 'European Federation', *Central European History* 1 14 ff.
[64] Lipgens, *Europa*, 309.

must be acknowledged that its impact upon winning the war was minute. The case of Tito's Partisans can in this instance serve as the exception which proves the rule. Maybe General de Gaulle was quoted correctly to have said, in an uncharacteristic moment of candour and self-mockery: 'Resistance was a bluff that came off.'[65] What matters of course is that it did come off in the wake of the massive effort of the Allied armies. But even if the *Résistance* was not effective strategically, if its impact was, as has been charged with some exaggeration, 'puny',[66] to the peoples it did give back pride, and to all men threatened by total dominion, hope.

In the German case the failure was more spectacular. Might we say that it was a 'non-bluff that did not come off'? The German Resistance proceeded from plot to plot, haltingly at first, headlong toward the end, culminating in the unsuccessful attempt on Hitler's life of 20 July 1944. The plotters' lack of popular support was compounded by the ignominy of failure. But in this instance, likewise, success should not be the ultimate measure of the intrinsic value of the attempt. Pastor Dietrich Bonhoeffer, whom I have cited before, during his 1942 meeting in Sweden with Bishop Bell of Chichester, one of the few friends in wartime England of the Germans of the Resistance, impressed upon him the function of Resistance as an act of penance: 'Christians do not wish to escape repentance, or chaos, if it is God's will to bring it upon us. We must take this judgement as Christians.'[67] In fact, most members of the *Widerstand* in the last and critical stages of the plot went ahead with their task, well knowing that it would not succeed.

But we must extend this negative stock-taking and register the fact that only in a few cases were the ideas, plans, and visions of the European Resistance—and I now include the *Widerstand*—translated into reality after the war. At any rate, I hasten to add, it is almost impossible to establish proper criteria for judging success or failure in this realm. The question is nevertheless often asked, especially in the Federal Republic: 'What if the Resistance had had its way after the war? Would it have made for a better world?' This will certainly remain an open question for ever.

In any case, I should like to suggest that resistance has a dimension transcending politics and purely practical calculations about success and failure. Those who resisted were human, all too human, many of them

[65] Foot, *Resistance*, 319.
[66] Ibid.
[67] Quoted in Ronald C. D. Jasper, *George Bell Bishop of Chichester* (London, 1967), 269.

timid all too long, many of them muddle-headed or foolhardy, few of them seasoned politicians or saints. But in their sacrificial act all of them, of whatever motive, persuasion, of whatever country, were one. What ultimately matters is that they were brave men, like Adam von Trott, whom we have all reason to remember proudly. It is the historian's task now to record that their death was not useless. Their stand against oppression has in the darkest of times reaffirmed the dignity of man. Their feats resemble those of poetry as W. H. Auden saw it when he wrote, 'Poetry makes nothing happen,' but added, 'it survives in the valley of its own making . . .'.[68]

[68] W. H. Auden, 'In Memory of W. B. Yeats'.

4

PROBLEMS OF THE GERMAN RESISTANCE[1]

KARL-DIETRICH BRACHER

I

MODERN times have seen the rise of the absolute state with its thorough organization of all spheres of life on strict bureaucratic and military lines; with this has gone an overriding concept of political obedience which has displaced the older tradition of the right to resist; it is still exerting its effect today on the relationship between authority and the subject. The revolutions in Britain, America, and France, of course, offered greater scope than ever before to the citizen's right of independent decision. Simultaneously, however, the principle of nationalism gained both in intensity and potency and as a result the newly won internal political freedoms gave way to increased state centralism.

This was particularly the case in Germany, where, after the failure of the 1848 revolution, liberalism, once so potent a force, withdrew in face of this development and postponed its internal political claims in favour of national unity and a strong foreign policy. For the sake of these advantages the semi-absolutist structure of the Bismarckian Reich was accepted; admittedly it was no police state, but nevertheless freedom of the subject was restricted to narrow limits indeed, as shown by the *Kulturkampf* and the anti-socialist law. In the minds of the citizenry the idea of freedom became applicable to external rather than internal politics, and as a result loyalty to the state became based on external successes; the capacity for constructive opposition, therefore, by which a parliamentary system based on the rule of law stands or falls, visibly withered. Little change took place when, after Bismarck's dismissal, confidence in the patriarchal state suffered its first setbacks. No more proof is required than the helplessness of the Reichstag in the years before and during the First World War and the small practical impact made by constructive political criticism prior to the revolution of 1918.

Herein lie many of the underlying reasons for the difficulties which were the hallmark of the career and the fate of the first German republic.

[1] For further readings see my books *The German Dictatorship* (Harmondsworth, 1980), 459, 553, and *The German Dilemma* (Weidenfeld & Nicolson, 1974), 104–31, from which this essay has been adapted and reproduced.

The republic was born not of some deliberate decision by the majority of the population to revise the relationship between rulers and ruled but of an unexpected defeat; it was a compromise emergency solution, unwillingly accepted. Its continuing crises and ultimate failure left their mark upon those who might have been prepared to resist the National Socialist dictatorship.

Naturally any statement to the effect that Hitler's dictatorship was a direct consequence of mass democracy and its system of equal universal suffrage is untenable. The fact remains that even with the most intensive propaganda campaign and under pressure of the most severe economic and political crisis, the National Socialists never succeeded in mobilizing more than one-third of the German electorate. It is equally significant, however, that the two-thirds who were opposed to National Socialism (or at least, not voting for it) on 30 January 1933 were incapable of producing any effective defence against the imposition of Hitler's dictatorship. They seemed to be in a state of stupor or self-deception, of susceptibility to illusion, opportunism, or readiness to accept *faits accomplis*; foreign countries reacted similarly when they accepted the Third Reich as partner in alliances or made concessions to it which they had refused to the Weimar Republic.

A major reason for this failure lay in the cloak of legality under which Hitler was nominated chancellor and was then able to progress from presidential coalition government to unrestricted dictatorship. Despite increasing loss of power by democratic and parliamentary agencies, when Hitler took the oath to the republican constitution an overwhelming number of powerful positions were still in the hands of non-Nazis. This was the case with most *Land* governments and with their police; it was so in the case of local governing bodies, the trades unions, and economic interest-groups; it was even the case in the Reichswehr, although there, as the position of the Reich president was soon to show, little reliance could be placed on formal positions of power to guarantee the exertion of any actual control over the new rulers or support of any resistance against them.

The history of the Weimar Republic had shown that a constitution, however carefully contrived, was of little assistance if attitudes of mind originating from the pre-democratic, patriarchal concept of the state persisted, giving rise to increasing suspicion of the republic and of democracy together with renewed yearnings for authoritarian leadership. Large sections of the populace, even if not caught up in the impetuous rise of the NSDAP during the closing years of the Republic, were not prepared for any form of resistance to the new regime; after all it proclaimed itself as the true German state as opposed to the feeble interlude of democracy. Like the majority of constitutional lawyers people overlooked the fact that the National Socialist concept of the state was diametrically opposed to one based on good order and the rule of law, that for Hitler and his *apparatchik*s the state was no more than an instrument for the pursuit of their racial and expansionist aims, that the National Socialist tyranny had developed parallel to and superimposed on the state, leaving the state as an empty façade. The party became the state and the state the party, as Waldemar Besson put it. The Third Reich's 'dual state', exemplified for Ernst Fraenkel in his book written in 1941 by the parallel existence of the judiciary and the concentration camp system, inevitably developed into the SS state and the SS empire. A victorious war was then to set the seal on what, as Eberhard Jäckel has shown, had been Hitler's *Weltanschauung* ever since 1923. Many ideologists of the state realized only too late that the Third Reich's ostensibly so legal and efficient totalitarian state was an illusion and that the reality was chaos, deliberately created by an arbitrary tyranny raised to the status of an ideology.

The fact, therefore, that the regime's *coup d'État* was camouflaged by a process of pseudo-legal seizure of power made the conditions for resistance and its prospects decidedly more difficult. Any attempt at constitutional protest or opposition according to the previous democratic rules of the game was stifled or suppressed by the security measures or terrorist methods of the new rulers, combined with a claim to legality put forward with considerable sophistication. Only two possibilities apparently remained: for the remnants of the persecuted groups, particularly the left wing, who escaped murder or the concentration camp and would neither resign themselves to the situation nor conform, disappearance into the 'underground', the logical step but a dangerous one; alternatively the attempt to acquire some influential position in the regime in order to organize resistance from within or at least have some possibility of exerting a restraining influence. This second course of action raised the whole problem of opposition by means of partial collaboration, a problem debated over and over again to the bitter end. Continuing loyalty to the authority of the state played its part here. People felt, too, that, if effective action was to be taken against the totalitarian system with its rigid controls, they must have influential contacts inside the machine. This, however, led to complicated situations covered by the ambiguous saying so frequently misused during those years for purposes of self-justification: that despite realization of the criminality of the regime one must remain in office in order to prevent something worse and exploit one's position to build up effective opposition from within.

This naturally applied with particular force to the civil servants, who, though severely intimidated, were left with some freedom of action owing to the lack of qualified Nazi personnel. The situation in the Reichswehr was similar, though there only a small number of officers were able to resist the attraction exerted by the community of interest between the military and the new rulers, who were so much more 'defence-minded' than the republican governments. Outside the state organization and in areas not so directly exposed to persecution and oppression the view generally held was that a certain degree of co-operation was acceptable in order to assist the—supposedly—good elements in the regime and protect one's own interests. This was the case over considerable areas of industrial and cultural life; it also applied to the ambiguous and highly debatable attitude of the churches to the National Socialist tyranny. In many cases it was a long time before the final decision between collaboration and resistance was taken—in contrast to those who were persecuted and outlawed in the early days and had long since chosen the first alternative, non-co-operation and the underground. All the more significant, therefore, was the fact that scattered groups were to be found working together consisting of men who, irrespective of their political or social background, stood out against terror and deception; they bore witness to the fact that, even in Germany and in the face of inhibiting tradition, the citizen's sense of moral and political responsibility had not been entirely eradicated. They looked back for support to the traditions of socialism, liberalism, and conservatism, too, to the religious and humanist view of mankind and the world. Their purpose was to break the paralysing yet intoxicating spell which the regime had cast over the whole life of Germany, but the decisive impulse for their actions came from a tortured conscience; this steeled them and gave them courage. Conscience had been aroused by the sight of fellow human beings being tortured, persecuted, and terrorized and would not be lulled by fashionable remarks such as Hugenberg's appalling comment with which he tried to console his following: 'When you use a plane, there will be shavings.'

There were stirrings of resistance frequently stemming from entirely individual decisions and cutting across all the old party and ideological boundaries. This gives the lie to the accusation so frequently made that resistance sprang solely from fear of loss of the war. Its origin lay in far deeper human and moral reactions, in a revolt of conscience produced in men of the most varied background by the brutality of the Nazi tyranny. The flame of resistance activity was lit by the fundamental and urgent duty to bring aid to the persecuted and dispossessed; men of similar views came together and exchanged information on what was really happening behind the smokescreen of official propaganda issuing from a regimented press and literature: finally they sought to organize contacts and build bridges to other people and groups in order to expand the basis of their charitable but at the same time oppositionist activity and finally proceed to political action. In the light of the historical and psychological inhibitions, of the profound problem posed by any decision to resist, the fact that there was resistance at all shows that more people had the courage to refuse to conform than had been the case in much of recent Germany history.

All this means that the decision to resist was one taken in terrible loneliness in the midst of a mass society. It entailed the permanent strain of suspicion, scepticism, and silence, danger for family and friends, isolation from the misguided majority of one's compatriots; to counter-balance all this there was no visible heroism, no halo of political glory. Here lay the difference between the German opposition and resistance movements in the occupied countries.

Above all, however, after the long decline of the bourgeois-democratic movement, the resistance could not trace its antecedents back to any political tradition, as could the political Left. The impulse to civil resistance sprang from the daily confrontation with the manifestations of dictatorship, from decisions of the individual conscience, and from the sense of right and wrong. This was the special feature of the German situation; historically one of the great services rendered by the resistance was that it brought some fluidity to the political groupings which had solidified under the Weimar Republic; its confrontation with the totalitarian regime led to a multiplicity of contacts across the frontiers of the old anti-fascist/socialist camp and the old parties or ideologies. Individual resistance circuits often formed spontaneously and they frequently included the most disparate social and political characters. This fact was of major significance for their subsequent wider co-operation, particularly after 1938.

II

Any discussion of our topic must start from the assumption that the decisive date of the German catastrophe is 1933, not 1939 or 1944–5.

This applies particularly to the problem of the resistance. Nazism's pseudo-legal seizure of power presented the old question of resistance to tyrannical domination in a new light. The difficulties to be overcome by an opposition to a state which, ostensibly legally, had come under control of a single party by means of emergency regulations, an enabling law, and plebiscites were quite different from those faced by the classic type of resistance to a violent *coup d'État* or open usurpation. The effect of the Nazi theory of the 'legal revolution' was to confuse people's minds and weaken all opposing forces. This both influenced the character and impaired the prospects of opposition to a regime which, in so short a space of time, had managed, by ostensibly democratic manœuvres, to impose an all-embracing tyranny upon the state and society. Moreover, the intellectual and social conditions of the time were not conducive to a definite expression of public opinion or the establishment of definite criteria by which the Nazi seizure of power might have been judged. In contrast to the West, when the modern German concept of the state was forming after 1848, the old tradition of the right of resistance to arbitrary authority was pushed into the background, swamped by the bureaucratic structure of the patriarchal state. This fact was highlighted by the problem of the oath sworn to Hitler.

The traditions of the German patriarchal state also provide some explanation of the mistaken ideas which exerted a paralysing effect on potential resistance to the Nazi seizure of power. For the Weimar politicians and parties the basic problem in 1933 was the extent to which they could influence or even control the new rulers by co-operating with them and the point at which opposition, in whatever form, became unavoidable. The problem is reflected, each in its own way, by the illusions of the trade union leaders, the wait-and-see attitude of the Social Democratic leaders, the bourgeois parties' fruitless attempts to adapt themselves, and the retreat of the German Communist Party. But in this way the moment was allowed to pass when action could have been taken from the old positions of power. Only as parties were disbanded and political positions lost did scattered centres of resistance form and even these were inhibited by the conviction that Hitler would quickly ruin himself economically and all that was necessary was to survive a short period of oppression.

A distinction must be made between the very different groups of opposition to Nazism or specific aspects of the Nazi regime; they emerged at different times with different methods and different aims.

Initial opposition came from the old political enemies of the Left and Centre, soon joined by disillusioned conservatives: then reinforcement came from the churches; there were also lone operators from officialdom or industry; finally in 1938 and again from 1942–3 the military became the focus of opposition planning and action. Evaluation is therefore both difficult and debatable because the criteria used may be very different—is the main emphasis to be placed on motives, prospects of success, or political purpose, for instance? Upon this will depend the estimate made of the left-wing, the bourgeois, the ecclesiastical, the conservative, and the military oppositon, their relationship to each other, and their anti-regime tactics.

The debate on the political, social, and ideological aspects of the resistance has so far suffered from the inhibitions resulting from the euphoria which its story produced. Hitherto there have been four main lines of argument. First, there are the two extreme views which place in a separate category (i) Communist resistance, which is labelled as 'treason', or (ii) that of the conservative and military faction, regarded as mere disagreement with the regime. This purely pro- or anti-Communist view of the resistance is historically unacceptable. From the point of view of the Nazi leaders with their claim to totalitarian domination both these forms of resistance represented a threat. Admittedly, once the regime was firmly in the saddle, it ultimately became evident that it could hardly be overthrown without the participation of the armed forces. This, however, is no justification for the next viewpoint (iii) namely that consideration can be confined to the military resistance; during the first half of the Third Reich's reign it barely existed and even after 1938 it can only be considered in conjunction with the political opposition.

Finally there is the view (iv) that the churches gave birth to a popular anti-Nazi movement and that the Catholic Church in particular was almost solid in opposition; this is as questionable as the opposing theory that a Communist, anti-Hitler mass movement existed. Opposition from the churches, highly important though it was, was a somewhat kaleidoscopic affair. Naturally it was a political factor, but only in a few cases did it rise above defence of its own positions and interests and turn into political resistance. On the other hand, criticism of the illusory ideas of the conservatives, including, for instance, the Kreisau Circle's constitutional drafts, objectively justified though it may be, overlooks the fact that at no stage in the Third Reich did a popular rising seem possible, that a *coup d'État* from above was ultimately attempted on 20 July 1944, and that this postulated contact with the official machine and parts of the establishment. This is not to deny that such attempts had their moral aspects and raised both spiritual and social problems.

III

At no point in time under the Third Reich, therefore, was there a unified resistance movement. Naturally at certain specific moments of crisis the multiplicity of political and intellectual forces which sooner or later either evaded or resisted the Nazi *Gleichschaltung* came close together; their general attitude and their plans, however, remained very different and their differences emerged in highly concrete terms after the end of the regime. Nevertheless the extent of internal German opposition in the pre-war period was far greater than the standard declamations of unity issued for foreign consumption would have one believe. Tens of thousands of the Nazis' political opponents were arrested; thousands were murdered for active opposition. Even disregarding the figures for those subjected to collective persecution outside the judicial system and in the concentration camps, the number of actual acts of opposition must be put far higher than Nazi propaganda would naturally have us believe; the Gestapo's secret surveillance reports give a very different picture. For many, of course, the step from non-conformism to civil disobedience and thence to active resistance was a large one but under totalitarian conditions the fact that it was taken at all constituted a political factor. The fact remains that although opposition came subsequently from the churches, the military, the bourgeoisie, and the conservatives, the first resistance was offered by those who (in Reichhardt's words) 'suffered first and most' from the Nazi regime's terror and whom the regime 'regarded as its most dangerous enemies—the organizations of the working-class movement'. It is particularly difficult to be specific on this; much is usually made of the persecution of the Left and the activities of the émigrés but sources are lacking for a comprehensive description of this widespread, faceless opposition. Under the conditions of totalitarian domination conspiratorial activity by left-wing resistance groups was confined to the darkness and anonymity of the underground; evidence is therefore meagre and far less revealing than the documents and plans produced by the bourgeois and conservative oppositon. Innumerable trials prove that this 'silent revolt' (as Weisenborn called it) was widespread and continuous; they often give a distorted picture, however, since the regimented Nazi 'legal system' invariably tried to present the accused as Marxist enemies; in many cases, moreover, action was confined to the concentration camp world, outside the judicial system.

The left-wing opposition was divided and too weak to initiate an active popular movement against National Socialism; this applied with even greater force, however, to all other social and political groupings. The left-wing parties, particularly the Communists, had a tradition of underground cell-building and organized resistance, but the others had no such foundation. As the years went by, increasing resignation became the order of the day in the socialist camp; expansion of the underground groups into a mass movement was obviously not succeeding and such opposition as there was was the work of individuals in contact with friends of similar persuasion; at the same time, however, the bourgeois camp was attempting to exploit its positions of power in state and society to capture starting-points for opposition and a change of regime. Initially three developments seemed to offer such starting points: partial resistance to *Gleichschaltung* by the churches, growing qualms in liberal and conservative circles about the true nature of the Nazi tyranny, and finally the criticism by disillusioned officers of Hitler's brinkmanship and gambling with war which really made its voice heard for the first time during the crisis of summer 1938.

The churches seemed most likely to provide a basis since they had been able to avoid complete *Gleichschaltung*. This was not universally the case, however. In later stages ecclesiastical opposition to the regime's claim to total domination was a political factor; it did not, however, imply *political* resistance in the strict sense of the word: its purpose was not to resist the Nazi 'authorities' as such but to preserve the churches' autonomy and educational freedom. In later years the fronts frequently became blurred; many compromises and concessions were made restricting active opposition even from the Confessional Church. The outcome was a sort of armistice; only a few travelled the whole logical road into the political resistance movement. Protestant resistance was handicapped by an inability to draw the political consequences, summarized as follows by Ernst Wolf, the theologian:

national and conservative prejudice, failure of liberalism as a way of life, an aversion to democracy (labelled 'Western Calvinism') which had become part of the confessional creed, support for the concept of the Reich combined with wishful thinking about a 'second' Lutheran-Protestant empire comprising the whole German nation against a background of the synthesis 'Throne (or nation) and altar', discontent with the Weimar Republic and with republicanism in general, anti-Marxist ideology turning communism into a bogey, the trauma of Versailles and, initially at least, defence of its own 'political Catholicism', finally an anti-Semitism which was only latent.

Such were the main Protestant prejudices and legacies from history; behind them stood the Lutheran doctrine of authority. In both churches only a minority was able to rise above 'ecclesiastical nationalism' and the myth of 'Reich, people, and Fatherland' inherited from the patriarchal state. Initially the churches had supported the Nazi state in principle; their subsequent opposition has been described by Wolf as 'an unwilling resistance movement'; it was a defensive reacation forced upon the Church to counter 'encroachments' but there were many relapses. As its conference in Stuttgart after the 1945 catastrophe the Evangelical Church acknowedged that it had been at fault—in contrast to the apologia with which other social groupings, including the Catholic Church, sought to escape their share of the blame.

So even in the churches only individual persons and groups, not the churches themselves, succeeded in reaching a clear-cut position in regard to resistance. This fact was further illustrated by the equivocal attitude of the churches to the Jewish question. Though they criticized the Aryan paragraphs of the Nuremberg Laws, this did not prevent the continuance of traditional anti-Semitism; both Protestants and Catholics protested against euthanasia but not against the Jewish policy; action was confined to individuals or measures of assistance. Only on rare occasions did the churches pluck up courage to issue general declarations. Major exceptions were the Confessional Church's resistance synods of 1934 and the memoranda and denunciations from the pulpit of the Council of Brethren (the most senior body in the Confessional Church) in subsequent years. In October 1943 the Prussian Confessional Synod, meeting in Breslau, openly disputed the government's right to proceed with its extermination policy, stating: 'The divine order knows no such terms as "exterminate", "liquidate", or "useless life". Extermination of men simply because they are relatives of a criminal, are old or mentally defective or members of a foreign race is not "a use of the sword which is the prerogative of authority".' This applied also to 'the existence to the Israelite people' and the statement ended with the words: 'We cannot allow our responsibility before God to be taken from us by our superiors.'

There were three problems with which even those closely involved in the churches' resistance could never come to terms. In the debate on the problem of the oath only a few managed to rise above the traditional loyalty to the state. On the subject of the war the churches mostly reverted to the 1914–8 attitude, giving patriotic duty and prayer for victory priority over many of their earlier scruples—even during the Sudeten crisis of 1938 the Lutheran bishops had kept their distance from the Confessional Church. The 'fight against Bolshevism' proved to be an argument which exerted an overwhelming influence even in circles inclined to be critical; similarly anti-Communism proved to be a bridge-builder between Protestant or Catholic thinking and National Socialist policy; it neutralized much of the opposition to the regime. Only a few followed Karl Barth in drawing the conclusion that war should be outlawed; equally only a few were found to agree with Dietrich Bonhoeffer's alternatives for the German people: 'either to hope for the defeat of their nation in order that Christian civilization might survive or to hope for victory entailing the destruction of our civilization'.

Many of the impulses and prejudices described here were common to both churches, though their manifestations may have been different. The difference lay, first, in the fact that, from the points of view both of organization and of dogma, Catholicism was far more highly organized and, second, was supranational in character. Nevertheless its relationship to National Socialism can by no means be clearly defined. Catholicism's surface unanimity did not prevent, in fact necessitated, early abandonment of its original opposition to Nazism in favour of acceptance of *faits accomplis*. Even more far-reaching tendencies to support the regime appeared. Admittedly the attempts to build an intellectual bridge between Catholicism and National Socialism never reached the strength or extent of the German Christian movement. The confrontation with National Socialism did not coincide with a severe structural crisis. For a time, however, the process of voluntary *Gleichschaltung* made considerable inroads even into the Catholic Church.

The main obstacle and limitation to Catholic opposition was the belief that it was both possible and essential to differentiate between loyalty to the state and criticism of the regime. Accordingly proclamations from bishops called emphatically for the return of the Saar to Germany and contributed to the Nazi victory in the plebiscite of January 1935; despite intervening experiences the process was repeated three years later with the significant declaration by Cardinal Innitzer of Vienna on the occasion of the Austrian *Anschluß*. A second factor was the support given to the war; this produced horrifying testimonies of a disposition to co-operate. Closely allied to this was the support and approval given to the anti-Bolshevist campaign. The theory that Germany was the bastion of Europe contributed largely to that ambivalent attitude which prevented Pacelli (Pope Pius XII from 1939) from issuing a public condemnation of National Socialism during the war. In his defence it is frequently emphasized that he failed to do so only out of concern for the precarious situation of the German churches and to avoid a crisis of conscience among the soldiery—but this merely highlights the problem further. Finally anti-Semitism was a traditional component of Catholic thinking and this prevented any divergence of principle from Nazi Jewish policy, though individuals were outspoken in their criticism of its methods and manifestations. Only the efforts of the last Papal Church initiated a change in the basically anti-Jewish attitude of the Catholic Church; recently it has become clear that this still persists today. For these reasons it is impossible to equate the Catholic Church with resistance. In contrast to the Evangelical Church there was no acknowledgement of guilt and no critical discussion in 1945; this fact has constituted a problem ever since the controversies over the Reich concordat of 1933 and the role of the Pope during the war. The fact remains, however, that misgivings about and opposition to the regime had a wider and more effective influence in the churches than elsewhere. The churches formed obstacles to the policy of ideological *Gleichschaltung* which even Hitler himself thought he could surmount only after a victorious end to the war. Pending that the regime relied on organizational restrictions, press bans, the arrest of prominent leaders such as Niemöller (in 1937), abolition of theological faculties, and intimidation of all pastors or priests who read from the pulpit the Confessional Church's declarations of opposition or pastoral letters. Underground activity and organization were the answer, particularly on the part of the divided Evangelical churches.

IV

The consolidation of the Nazi regime confined resistance both by the socialists and the churches within strict limits. There was no hope of a popular rising and neither were strong enough to bring about a change on their own; the only possibility was to gain contact with men in powerful positions in society, the state, or the armed forces and so bring some influence to bear on the Third Reich's political and military decisions. It became clear that under a totalitarian regime a popular opposition movement had little prospect of starting or doing anything effective and that the masses were ill suited to illegality and resistance. This meant that the rise of anti-totalitarian opinion among officialdom and the military, neither of which was basically democratically minded, acquired increased importance. Outstanding in this respect was the tireless activity of Carl Goerdeler—the ex-German-Nationalist and ex-Burgomaster of Leipzig—in the years that followed. He pursued three main lines of action: he attempted to influence the nation's leaders by memoranda; he established contact between the various circles of the emergent bourgeois and conservative opposition; and he brought influence to bear on the bureaucracy. Finally, as the conviction began to grow that revolution from below was out of the question and that only a *coup d'État* from above was possible, the resistance began to turn increasingly to the army.

The Wehrmacht was anything but prepared for this role. Contrary to the hopes of many conservatives it had accepted without protest the seizure of power, the murder of Generals Schleicher and Bredow, and the oath to Hitler. Though its attitude was based on illusion it nevertheless helped the Nazi regime to establish its tyranny. At the root of this was Seeckt's ideal of the 'non-political soldier', the background to which was in fact anti-democratic—Hitler had succeeded in pacifying the old school of officers by guaranteeing them their autonomy. The generally held view, moreover, was that the military interest coincided with that of the Nazis; a policy of rearmament and removal of the Versailles restrictions were aims for the sake of which officers were prepared to accept many of the Nazi practices as mere aberrations. After its débâcle in 1934 the Wehrmacht was no longer in a position to act as a power group of its own. Even military resistance, therefore, could only be a partial movement, the effort of a minority which might temporarily possess itself of important positions and contacts but could never succeed in exerting adequate influence at the top level. This determined both its form and its limited possibilities.

Events proved that the German military tradition offered no adequate foundation for political resistance; military resistance remained confined to individual initiatives on the part of independent-minded officers. Its root lay, not in the traditions of the Reichswehr, but in decisions of conscience on the part of individuals. For far too many officers tradition was used as a reason or an excuse for evading the crisis of conscience on the pretext of their duty to obey and refusing to participate in resistance to the end. Nevertheless in the summer of 1938 the first military resistance group to produce definite plans formed around Beck; for the first time almost all shades of political opinion were represented. The conspirators were in contact with the Social Democrats, trade union leaders, senior civil servants, and the Wehrmacht's semi-civilian secret service; they planned for a change of regime at the moment of the anticipated military crisis. From this time the centre of active conspiracy was to be found in a circle which formed inside the *Abwehr* (Intelligence) Division of OKW (*Oberkommando der Wehrmacht*—High Command of the Armed Forces). They were the most closely in touch with the true situation; the driving forces were Colonel (as he then was) Hans Oster and later Hans von Dohnanyi, a High Court attorney (*Reichsgerichtsrat*). The central feature of the plan was the arrest of Hitler the moment he issued the order for war, which, it was anticipated, would be followed by declarations of war by the Western powers. It was calculated that the German people would be so taken aback that the enterprise would have wide support and so the threat of civil war would be avoided. The resisters had reason to hope that, once the criminal and catastrophic nature of Hitler's policy had been made plain to all, even the disciplined bourgeoisie and the military would support them and that refusal to obey would not appear as sabotage and treachery. As a warning background to these expectations, however, stood the experience of November 1918.

These plans were designed to avoid two dangers: a civil war, the outcome of which was uncertain in view of the power of the Nazi Party, and a stab-in-the-back legend in reverse—a future new regime might have been saddled with the accusation that the army and opposition had stabbed Hitler in the back at the moment of victory. The execration of which the action of 20 July 1944 was later the target shows how well founded these fears were, although, even had it succeeded, it would merely have brought to an end a war long since lost. Hitler's triumph at the Munich Conference, however, cut the ground from under the feet of these plans. In the next three years the regime went from victory to victory. As its prestige grew the ranks of the opposition thinned and action against the regime had practically no hope of success; a putsch would inevitably have brought the consequences the conspirators feared—civil war and a stab-in-the-back legend. Any future action was now dependent on military participation which was problematical and hesitant anyway; the failure of summer 1938 introduced additional complications. A further factor was the increasingly strict measures for the security of Hitler's person taken during the war.

V

At this point the moral and political problem raised its head: when and how the opposition could and should have recourse to force. The churches were not alone in raising objections of principle and, with a few exceptions, opposed the use of force, including the killing of Hitler. Many in the bourgeois and conservative opposition, from Goerdeler to the Kreisau Circle, were undecided on this problem and were content to leave an attempt at assassination to the military, the majority of whom clung to their oath and their duty to obey orders. This constituted a major problem not only in the preparation of the *coup d'État* but also in obtaining agreement between the various opposition groups; in the end it decisively prejudiced the attempted *coup* itself. To the very end far too much confidence was placed in subterfuge and surprise. The basic pattern of the September 1938 plan was followed: first ensure military support for the *coup*, which was, if possible, to be bloodless, then win over the populace by proclamations and dissemination of reports about the criminal nature and catastrophic policy of the Hitler regime.

War both complicated and assisted the opposition's task. On the one hand it became increasingly difficult to differentiate between National Socialism and Germany; the appeal to patriotism was stronger than misgivings about the regime. Further factors were the wartime increases in regimentation and controls and the general acceleration in the tempo of life. On the other hand, war entailed a greater degree of improvisation and pragmatism; it also tended to make the hierarchical structure, both in the civilian and military fields, more fluid and accessible to outside influences and personalities; this obviously helped the resistance to organize itself and expand. Most important of all, however, war meant a sudden and violent increase in the weight carried by the Wehrmacht, which, despite its previous feeble reaction, had kept its distance from the Party and, above all, the SS. Moreover, there were now numerous civilian opponents of the regime in military posts. Covered by Admiral Canaris, Oster had been recruiting men like Dohnanyi and Dietrich Bonhoeffer into the *Abwehr* from the outbreak of war.

In the situation of 1939–40, before the war had spread, these men maintained contact with the Allies; Oster even attempted to give one final proof of the opposition's sincerity by passing to Holland the dates for the German attacks on Scandinavia and France. Even today many critics of the opposition still decry this effort, like all foreign contacts, as 'treason'; it has even been used as a pretext for a new stab-in-the-back legend. It was in fact an expression of Oster's uncompromising opposition and his determination to do his utmost to end the war and overthrow the regime. Under a dictatorship the distinction between treason against the government and treason against the country is blurred in any case, but Oster's action was justified by his knowledge that Hitler was about to assault five neutral countries whose inviolability he had explicitly guaranteed. Both politically and morally Oster had good reasons; he was only too well informed about the unscrupulous nature of Nazi preparations for aggression. If treason against the country implies an intention to damage one's own country, then, even in the case of this exceptional step, right was on the side of a man who was striving with all the resources at his disposal against breach of treaties and violation of law. Here was a state, based on injustice and disregarding all its obligations both to its own citizens and the outside world; in such circumstances it may be thought that greater right is on the side of treason against the country and breach of an oath. Since the outbreak of war confidence in the German opposition had been shattered and Oster's action was a determined and desperate attempt to re-establish it. His efforts were of no avail, since his warnings were not taken seriously and the military efficiency of the German operations led to an unexpectedly rapid and complete victory in the West.

Hitler's new triumph changed the situation fundamentally. The German victory over France implied the most severe defeat for the opposition. The resistance movement now entered upon a period of severe trial, for it had to maintain itself in face of the enthusiasm generated by a victorious dictatorship. All previous contacts with the West had vanished, as had hopes for a quick end to the war and revolt at home. The opposition was isolated, without prospect of winning popular support; almost its sole remaining strength lay in the moral and humanist basis of its existence; there seemed to be no grounds even for thinking of any concrete external action. The degree of continuity, therefore, with which the opposition organized and expanded itself, seems all the more remarkable. The fact that it did so is the answer to those who later maintained that German resistance sprang merely from fear of defeat, from an eleventh-hour panic. This may be true of certain of its military champions. It is certainly not true of those who, at the time of the Third Reich's greatest victories, continued to carry the burden of the perilous battle against Hitler and his apparently invincible regime.

Fresh plans for the overthrow of the regime could not follow the previous lines; even less than in 1940 could any help be expected from the military, now intoxicated by victory and in many cases promoted and bemedalled. Vast members of Hitler disciples had been accepted into the officer corps at the junior and medium levels and this inevitably affected the generals' readiness to act. Moreover, as the power of the SS grew, the regime's security measures were continually being tightened. Since the start of the Russian campaign Hitler himself had been living almost exclusively in the bunkers of his hermetically sealed headquarters in East Prussia or the Ukraine; planning for any future assassination attempt was a matter of the utmost difficulty. This was doubly serious since it became increasingly clear that, under wartime conditions and in view of the personal power of Hitler, the murder of the dictator was the *condicio sine qua non* for any change of regime.

VI

From other chapters in this book we know of the important role Adam von Trott zu Solz was playing with respect to the foreign contacts of the German opposition. We know what happened on 20 July 1944 and we know the frightful consequences of that failure—the regime claimed more victims than ever before. This meant, of course, that National Socialism's full responsibility for the catastrophe could not be obscured by some new stab-in-the-back legend, although such a legend has been pertinaciously cultivated by widespread neo-Nazi publicity of which the NPD's utterances and the *Nationalzeitung* form part. The outcome was that, in an atmosphere of spellbound terror, the mass of the people followed the regime to the bitter end, even when that regime's leader issued from his Berlin bunker his senseless orders for resistance to the end and scorched earth. So the question with which a *coup d'État* would have confronted the German people was shelved and remained unanswered: what had the people to say concerning the crimes committed in their name? The reckoning with the Nazi criminals which Stauffenberg demanded should be conducted by Germans never took place; in Ehler's words it remained 'frozen in the realm of the past, irrecoverable'.

The much discussed 'conquest of the past', however, was not the only failure. There is no certainty regarding the present-day significance to be attributed to an experience which most people have either pushed into the back of their minds, forgotten, or consigned to ancient history. Not unnaturally, on the surface the second German democracy traces its origin back to the resistance. Reference is made to the 'other Germany' as an argument against the verdict of the victors or the equation of Germany with National Socialism. At the same time, however, it cannot be denied that this 'other Germany' was a small minority, ferociously hounded and destroyed, never acknowledged by the vast majority of the population.

Democracy stands or falls by recognition of the existence and validity of opposition. Opposition is a permanent duty, not a natural state of affairs. Karl Jaspers once said that opposition was 'contrary to human nature; man wishes to live at one with his environment, to vote Yes in concert'. Nevertheless it is precisely what the logic of democracy requires. Because everyone can be involved and therefore carry a share of the responsibility, democracy needs a knowledgeable, critical public and precisely for this reason that public should not only tolerate but acknowledge, and if necessary protect, minorities which think differently. The dictatorial state, not being based on the rule of law, necessarily carries the odium of illegality and use of force; there can be no comparison between resistance to it and non-parliamentary opposition to the system as a whole in a democratic constitutional state. The argument generally used to justify such action is that political domination is tantamount to use of force and that therefore use of force against it is legitimate. This, however, bears no relation to the reason which justifies resistance to dictatorship—the demand for re-establishment of constitutional conditions and the safeguarding of human rights.

Here the fundamental distinction emerges: resistance is undertaken to safeguard the rights of man when the constitution of a state based on the rule of law has been violated; revolution designed to overthrow the state is embarked upon in the name of perfectionist utopian visions of the future and without regard to the sacrifices involved. Obviously there can be various intermediate situations; even in the case of the anti-Hitler resistance there was no question of mere reversion to the pre-dictatorial or even pre-republican conditions; this fact placed a question-mark against the alternative drafts produced by the Goerdeler group, which was both constitutionally and socially conservative. The object was to re-establish a liberal social state based on the rule of law and to put an end to arbitrary misuse of the law in the service of superhuman or inhuman purposes. The argument may be adduced that a right of revolution exists and that, had it not existed, modern democracy would never have been born. On this subject, however, Germany, the country of frustrated and mistimed revolutions, has much leeway to make up.

The bankruptcy of 1933 and the failure of the resistance under the Third Reich, however, contain a lesson: it is that a civilized political democracy is dependent upon the right of opposition being established both morally and by the constitution. In contrast to the western democracies no generally accepted tradition of the right to resist continued in Germany when the patriarchal state was established by revolution from above after 1848. Neither the Weimar Republic nor our second democracy has been able to build on a solid, generally accepted, concept of the state on the part of its citizens. The Third Reich has therefore left us a dual legacy: the tendency either to take refuge in the comfortable security of the patriarchal state or to erupt into extremist movements opposed to the system as a whole; as a result we are in danger of reverting to the extreme polarizations of the Weimar Republic.

Both these trends show that we have not yet learnt the painful lesson which the failure of the Weimar Republic and the sorry history of the resistance should have taught us. The title of a recently published paperback much oversimplifies the problem; it is: *German Resistance—Progess or Reaction?* Certain Germans refused to be seduced; they resisted the call of opportunism and unthinking intoxication with power and victory; they fought alone, accepting persecution and death. Their sacrifice is not to be judged on the basis of the fashionable capitalist—socialist alternative nor of controversial moral and theological principle. The overriding aspect of the resistance seems to me to be that it demonstrated the old conviction, long abandoned in Germany, that there must always be two sides to politics: on the one hand construction and consolidation of a social order, on the other, resistance to biased and unjust exercise of power. Under certain systems of government there is no room for a difference of opinion of this nature; their aims may be the establishment of a classless society, maintenance of order in an established, patriarchal state, or the victory of a militant ideology which considers itself the sole repository of truth. Whatever its purpose, however, such a system of rule violates the basic law of any politics worthy of the human race.

The dead of the resistance are martyrs, witnesses to this conviction which has formed the basis of western politics ever since the days of Greek democracy and revolt against the power of tyrants. They were carrying on this international tradition renounced by Germany with her anti-western cult of the state. This is all the more remarkable in that they were largely men and women who had been reared as prisoners of that traditional, disastrous, and specifically German concept of the state which called for obedience, devotion to duty, authority, and national power. They were not naturally inclined to revolutionary views or revolutionary action but they 'put back into the German vocabulary' (in the words of W. Hennis) the words 'resistance' and 'tyrannicide'; they were able to rise above the German metaphysic of the state, the belief in the state as the be-all and end-all. The fact that they did so consitutes a milestone in the history of our political consciousness.

The basic lesson of the German resistance is that the state authority and the nation should no longer be regarded as absolutes, that the primary objects of the citizen's loyalty should be a standard of values transcending the state and a form of political process worthy of the dignity of man, that the overriding consideration should be, not the state or some oath or some order of society as such but an alert constitutional consciousness, a democratic consciousness. Herein lies the great justification for a resistance which has been vilified as illegal and treacherous. A critical political sense must prevent the resistance being pushed into the background and written off as historically irrelevant. At the same time, however, if the opposition in a pluralist democracy, whether on the extreme Right or the extreme Left, carries the banner of some one-sided, perfectionist utopia, we must resist the temptation to place a halo round its head and call it resistance.

Thomas Mann once called politics an attribute of humanity. The significance of the 1933–45 resistance, into which such loyal citizens as Trott zu Solz and Ludwig Beck found themselves forced, cannot be grasped if its political and moral aspects are considered separately and in isolation. Its significance lies in the close connection between the two—the one has often been played off against the other in Germany's more recent history and in interpretations of the anti-Hitler resistance. We are faced once again with the old question of the relationship between the law, justice, and force. The answer given by the Germans of 20 July was not that force was justified, as the superficial saying goes; it was that they were prepared to accept the odium and the sacrifice entailed by the use of force in order to re-establish the rule of law and justice in the face of inhuman criminal tyranny. Their legacy is that force should no longer be necessary since freedom, the rule of law, and peace have become established as the main pillars and purpose of any state constitution.

5

THE THIRD REICH AND THE GERMAN PEOPLE

MARTIN BROSZAT

IT was as a young Rhodes Scholar at Balliol College that Adam von Trott zu Solz, one of the most courageous men in the German resistance against Hitler, established contact with David Astor, Richard Crossman, Stafford Cripps, and many other persons later influential in the political and public life of Great Britain. Before and during the war, he tried unremittingly to use these contacts to obtain assurances from London and Washington as to what their response would be were Hitler to be overthrown. But in spite of all support from some of his British friends, the British government, especially the Foreign Secretaries Halifax and Eden, remained reluctant. Among other reasons there was one essential cause for suspicion: Adam von Trott, although he strongly condemned the Hitler regime from early on, was and remained a passionate German patriot with certain traditional national convictions. Like other representatives of the German resistance in the officers' corps or the diplomatic service, von Trott assumed that after an overthrow of Hitler Germany should not give up all the territorial acquisitions gained during Hitler's reign.

This assessment has, I think, a much more than personal significance. It means that the national goals of the traditional German political and social élite and the goals of Hitler's foreign policy were partly congruent for a long time, despite extremely different views concerning the means by which they should be pursued. And this, furthermore, was one of the major factors which gave Hitler such widespread support in Germany up till the first stage of the Second World War, support not only from the German middle class but also from the old political élite, from those sharing influence and power during the Nazi regime in the armed forces and the diplomatic and bureaucratic service, and even from most of the leading representatives of the Protestant and Catholic Churches. The power-cartel of the Third Reich, otherwise divided by different moral and political perspectives, had one main uniting element: their desire for a full national recovery after the loss of the First World War, a revision of the Treaty of Versailles, the incorporation of German nationals living beyond the German borders, and the re-establishment of a more powerful Germany and a bigger German role in Europe. The popularity of Nazism cannot be understood without the historical interconnection between Hitler's foreign policy successes and the national expectations of the traditional German upper class—expectations which had been shattered as a result of defeat in the First World War. The broad nationalistic mood of the common people, from which Nazism so greatly benefited both before and after 1933, had not come up from the people themselves but had been implanted into their thinking and feeling in the preceding decades by the leaders of opinion in the traditional German élite in government and society, in universities and schools, by intellectuals and priests, and, last but not least, by the patriotic organizations sponsored by these élites both pre-war and post-war. The pattern of this traditional national thinking contained the concept of a German *Machtstaat*, of a 'Greater Germany', and of German predominance in Middle and East Europe, but it contained also specific manifestations of German national ideology, which since the Romantic era suggested a separate and unique course of German national self-realization in contrast to western civilization and democracy. The dream of a special German course in history, the dream of the Third Reich, had much older sources and much deeper roots than Nazism. And a specific historical constellation played an important role here. Massive population growth since the late but swift formation of a German industrial society, and the spread of national feeling and political consciousness among the masses in the second half of the nineteenth century, came together at a time when the traditional political class in Germany, mainly in Prussia, had more and more turned away from the older forms of pre-national liberalism and conservatism and, sparked by the euphoria caused by Bismarck's founding of the Reich as well as by the dynamic development of the German industry and trade potential, were swept along by the rising and often rampant tide of a national *Machtstaat* philosophy.

This philosophy and the prestige of the conservative élite received a heavy blow with the defeat in World War I. But the peace treaty of Versailles, perceived as a grave injustice even by the German Democrats and Social Democrats, hindered in the following years the formulation of a new and convincing foreign policy for the Weimar Republic and kept the traditional national ideas alive. The solidarity of all German political parties on the principal question of a revision policy, pratically forced upon them by the Treaty of Versailles, hardly permitted a candid criticism of the German pre-war *Machtstaat* policy.

On the contrary, the frustration of national goals and the inability to change the situation gave rise to a kind of introverted *völkisch* nationalism. Vague, but passionately cultivated ideas of German *Volkstum* and *Volk* arose and preoccupied German national thinking to a wide extent. National Socialism arose in the atmosphere of this *völkisch* ideology. In this context, during the mid 1920s, Hitler formulated his long-range racial goals and his concept of *Lebensraum*, which revived the dream of a great and powerful German Empire. But now, based on the supposed racial superiority of the German people, this dream was redefined and transformed, more unscrupulously than ever before, into the goal of conquering and establishing an empire of the German master race in the east.

The partial coincidence of Hitler's goals with the national and hegemonic notions of the old political élite clearly foreshadowed the alliance between Hitler and the old conservative ruling class concluded in 1933, but also the later opposition from conservative circles, when it became apparent (as it did by 1938 at the latest) that Hitler's methods were irreconcilable with their feeling for national responsibility. This feeling had remained alive in at least a part of the old political élite. But the vast majority of them had already been corrupted by Hitler's sensational successes in foreign policy. The loss in prestige and credible capacity for leadership of the old political class, which had already been apparent in the areas of social and constitutional politics long before 1914, revealed itself under Hitler in the field of foreign and military policy as well. The political success of Nazism was not only a result of breakdown of the democratic parties of the Weimar Republic. It was also the result of the progressive loss of credibility and influence of the old political class. Their self-deceptive weakness became especially apparent in the coalition cabinet in which Hitler was named Chancellor.

By stressing this connection between the rise of Nazism and the breakdown in the leadership capabilities of the traditional German political class, I have suggested a pattern of historical interpretation of Nazism, which in what follows will be illustrated further. Limitations of space allow me to make only some general remarks.

In the wide range of literature on this topic, by German and non-German authors alike, growing doubts have been raised as to whether one can speak at all of a special German course in modern history. Comparative studies demonstrate the difficulty of distinguishing specifically German forms of imperialism during the pre-Hitler era; they show that the rise of vitalism, social Darwinism, or cultural pessimism, upon which Nazism based itself, was not restricted to Germany. Even the extremely fast pace of industrialization in Germany during the nineteenth century within the political framework of a constitutional but at the same time military monarchy and the resulting discrepancy between an industrialized civilization and a semi-feudal and pre-industrial political and social culture may be regarded as a crisis of modernization that is not specifically German. Doubts as to whether all these and other factors are really unique in the development of pre-Hitler Germany are the more well founded to the extent that the simultaneity of fascist movements in many European countries in the inter-war period indicates the existence of general causes. An examination is required of the epochal character of the crisis of the liberal democratic political systems which ensued after the First World War and the challenge of the Bolshevik revolution. Nevertheless, the fact that Nazism in Germany was by far the most complete, the most energetic, and the most violent form of fascist government calls for an explanation on the basis of German history.

But it has also become evident that the historian has only rather weak instruments at his disposal when attempting explanations of this kind, when he wants to measure more exactly the significance of various historical interrelationships. The further he reaches back into a period of time with totally different conditions, the more questionable becomes his claim to causal explanations. It is therefore, I believe, no oversight but rather a wise self-limitation on historians' part that their debate about the specific historical causes of the Third Reich has for a long time concentrated mainly on the relatively short term, on the history of Germany as a nation-state and its industrialization since the nineteenth century. And I myself believe that the cause for the spread and rise of Nazism is to be found above all in the turbulent maelstrom or crisis, simultaneously national, political, social-economic, and cultural in nature, occurring in Germany after the First World War.

Apart from this question of historical preconditions, our topic raises a wide range of further issues. An investigation of the popular basis of the Nazi regime leads into the sociology of Germany in the pre-Hitler era, into the broad diversity of political traditions, provinces, and social and cultural milieux, some of them more resistant to Nazism, and others more favourable toward it. Such an investigation would also require studies of the effectiveness of Nazi propaganda and the organizational structure of the totalitarian rule created by the Nazis. In contrast to the heritage of German national thinking, of which mention was made above, it is most important to explain the underlying social processes and socio-psychological motivations which contributed to the success of Nazism below the surface of political radicalization.

The overwhelming success of Nazism, which in the final stages of the Weimar Republic reached its peak in July 1932 (when the Nazis obtained 37.9 per cent of the vote), cannot be considered only from the viewpoint of the basic national mood and the ideology of right-wing radicalism. The core of the Nazi movement, with its militant SA storm-troopers, with leaders like Hitler, Göring, Geobbels, Röhm, and Himmler at the top, consisted of a group of radicalized, political desperadoes, who in the name of an aggressive and utopian ideology expressed over and over again their deep contempt for humanitarian values, for the morality of educated bourgeois society, and for the rule of law. But what the historians of today consider to be the most inhuman of the basic components of Hitler's ideology and programme—the fanatical hatred toward the Jews and the goal of a master-race empire in the east—were not the determining factors in the tremendous success of the Hitler movement immediately before 1933. Neither were they in the forefront of Nazi propaganda at the time. In July 1932, over 13 million Germans voted for Hitler, not because of his racialist ideology, but rather in spite of it. At the core of Nazi propaganda at that time, besides the fight against Marxism, was above all the highly effective promise to overcome the splitting of soical classes, religious groups, and political parties in a so-called *Volksgemeinschaft*. What attracted most followers of the Nazis was their seeming efficiency and credible promise of a more stable government and a stronger leadership, demonstrated by the organizational strength and activity of the Nazi movement. Its almost total absorption of the middle-class and conservative right-wing parties, with the exception of the Catholic Centre party, reflected the small public response to the efforts of men like Brüning, Papen, and Hugenberg, who tried to bring about a more or less authoritarian change of the democratic system of the Weimar Republic. Hitler's ability to attract voters from all classes of society, the populistic appeal of Nazi propaganda, and the broad basis of the Nazi movement with the lower middle class in both urban and rural areas, all are symptoms of a strong need among these strata for a higher degree of social equality and mobility, which the Nazi party knew how to inflame and to use for its advantage. The masses who turned to Hitler were seeking a new orientation, a new authority, and a new commitment, but something which was more popular and modern than Papen's Cabinet of Barons could offer them. The Nazi party's mass contingent from the rural population, as well as that from the old and new middle class, was not only determined by ideologies and the radicalization of conflicts of interest during the economic crisis. It was also motivated by strong demands within the rural and middle-class groups for more social participation. The appeal of the Nazis meant also the discarding of social patriarchalism in middle-class and rural associations in the provincial society of small towns and villages. The success of the Nazis in these social spheres also expressed the desire for more proximity to the grassroot problems of the common people, for more social rejuvenation, which had been lacking the Weimar Republic owing to the perseverance of the old élite. Along with the engine of revitalization and appeal to youth, which the Nazi movement developed, there was also a measure of justified criticism of the immobility within German society. This mood of social unrest and protest, strongest among the younger generation, was a motor for the Nazi movement hardly less significant than the backward-looking romanticism of its political ideology. The most effective element of Nazi propaganda, the *Volksgemeinschaft* slogan, was not only an unreal utopia, not only a recourse to the feudal structure of society. It was also a call to overcome the relics of the pre-industrial and pre-democratic social hierarchy and norms, an appeal for the forming of a modern and mobile mass society.

It was for this reason, and not because of the anti-Semitic or social-conservative elements in the twenty-five-point programme adopted in 1920, that the Nazi party appealed to such a wide range of young people in the rural and urban population, to students, young engineers, and technicians. Without the potential of these aspiring and efficient social forces from the middle class, without their strength to stand up against the older generation and break out of the older, time-honoured traditions in family, church, school, and profession, the activism of the Nazi movement can hardly be explained. The same holds true for the energy of the new political élite, who got their chance during the Third Reich. It was during the rise of Nazism to a mass-movement and as a result of opportunistic recruitment to the Nazi party in the first months of the Third Reich that the regime was able to call upon not only the old party veterans, but also on a new and, in many respects, extraordinarily competent élite—the likes of Reinhard Heydrich, Wilhelm Stuckart, or Albert Speer. Ideological motivation amongst these aspiring young experts, most of them capable technologists of power, was usually rather weak; far stronger was their energetic and often unscrupulous ambition to rise up, which often could be realized much more easily under the unconventional conditions of the Nazi regime than in the encrusted society of the Weimar Republic.

In this context mention should also be made of the Nazi cultivation of the theory of natural selection through the continuous struggle for survival of the fittest, and the many forms of organized contest which were set in motion during the Third Reich. Apart from the effects of ideological indoctrination and organizational control, which the Nazi regime implemented, social motivations could be used to a large extent to transform the old bourgeois society into a morally liberated, nationally oriented, dynamic, efficient, and productive society. It was not only the Hitler myth and the race-ideology that induced a dynamic mobilization of large portions of society, especially among the younger generation. It was also the feeling, constantly kept alive and spurred on by propaganda, that one was living in a future-oriented, more open and mobile society than before 1933. The political system of the Third Reich, with its mass-organizations, further undermined the already fragile prestige of the old opinion-leaders and traditional élites on all levels of society. The village schoolteacher received support from the party in his need to free himself from the spiritual superiority of the village priest and the traditional practice of clerical school inspection. The *Hitler-Jugend* and the *Bund Deutscher Mädel* carried the principles of juvenile autonomy without adult supervision and greater freedom of movement among juveniles to the most remote areas in the country. The young officers in the army, who had been educated in the Hitler Youth, brought adoration for Hitler as well as some fresh blood and a new trend towards equality into the old caste spirit of the officers' corps.

The social propaganda and policy of the Nazi Labour Front, too, was able to produce effectively the impression of a breakdown of the differences of social classes, even though the rights of self-determination and collective partnership of the working class had been abolished. When one considers the social mobilization and equalization, at least in the consciousness of many people of the lower middle or working class, it is difficult to dispute the thesis of the modernizing effect of Nazism. Moreover, there was a connection between this increased social mobilization and the enormous readiness of the people for energetic action, which the regime knew how to stimulate. Even foreign observers, with all their criticism of the political system of the Third Reich, noted the vitality of German society, the modernity and innovativeness which could be felt in Germany in the years from 1935 to

1938. A leading English newspaper benevolently described the social and political system of the Third Reich in 1938 as a totalitarian democracy. The Weimar Republic constitutionally guaranteed the legal claim to equality and democratic rights; in reality, however, it could not achieve much of the substance of equalization, mobility, or social-political participation. On the other hand, Hitler offered for many—at the expense of personal freedom—greater opportunities for a socio-psychologically motivated expansion of personal existence and for a suggestive psychological integration. Up till that point in German history, there had been no successful democratic revolution that had given the large middle class a chance to feel as dominant. But now, under Hitler's dictatorship, the middle class experienced a kind of compensatory pseudo-revolution in the form of plebiscitarian and totalitarian participation. The approval of the Nazi regime by the majority of the German population after the successful battle against mass unemployment and Hitler's foreign policy successes had an essential basis too in the socio-psychological mood of the people, which had resulted from the long-standing authoritarian and patriarchal mould of German society.

Related to the populist appeal of Nazism was also the effectiveness of the Hitler halo. The expectation of a new leader who was capable of uniting and lifting the nation out of confusion, disruption, and depression had already been shaped by the political Right in the late Wilhelmine era, for example by Heinrich Class, the leader of the Pan-German League, in his book with the title *Wenn ich der Kaiser wär'*. In this pamphlet the vision had already been planted of a leader who, emerging from the ranks of the people, expressing the emotions of the people, and speaking the language of the people, stood up against the existing reality of the authoritarian monarchy. Hitler was to fill this role and these expectations. His demagogic skill and his effectiveness as a speaker to the masses rested mainly on the fact that he, more than anyone else in post-war Germany, could, authentically and credibly, formulate, emphasize, and captivate the feeling of the masses—the resentment, the depression, and the longing for redemption among a large portion of the nationally minded bourgeoisie, which resulted from the national, political, and social lack of orientation after World War One. The aura of a missionary-like, political revivalist quickly won him the undisputed leadership in the party, which since the beginning of the thirties became known more and more as the Hitler movement. First presented as *Volkskanzler*, then as *Volksführer*, by all the organs of Nazi propaganda,

Hitler and his charismatic emanation soon after 1933 became the superior integrative power of the Nazi regime, far more effective than the Nazi *Weltanschauung*. The *Volksempfänger*, a small, cheap radio introduced in 1933, soon transmitted into almost every household public appearances, meetings, and speeches by the Führer; in this way, large portions of the population participated emotionally in his national policy and demagogic self-projection. By 1935 prestigious conservative authorities in constitutional law saw in the embodiment of the will of the people through Hitler the real legitimation of the *Führerstaat*. Actually, the plebiscitarian response to Hitler substantially contributed to strengthening his real power as absolute Führer in relation to those who shared decision-making influence in the armed forces, bureaucracy, and economy. The popular belief in Hitler's infallibility, which became stronger and stronger with his triumphs in domestic and foreign policy, was—so to speak—always present when ministers, generals, diplomats, or economic leaders had to deal with him. Hardly anyone approached him in an unaffected manner any more; many tended to praise even the strangest of Hitler's ideas as the prophetic view of a political genius. The aura of the charismatic *Volksführer* which surrounded him more and more created a growing distance from Hitler, even at the top level of state and party administration. This separated him from, and lifted him above, the rest of his government, thus producing the basic psychological requirement for the absolute rule of the Führer. The final form of this was that Hitler, from the remoteness of his residence at Berchtesgaden or his military headquarters, often informed his ministers or party leaders of his supreme will only through submissive secretaries, adjutants, or personal servants. The growing inaccessibility of Hitler sharpened the competition among his subordinates to obtain directives, endorsements, and authorizations of power from him and also caused arbitrary interpretations as to what the will of the Führer really was. By managing this, Martin Bormann, Hitler's secretary, in the end became the nearly omnipotent head of the party chancellery.All this marked the court-like state of the Third Reich during the final stage of the Nazi regime. These structural conditions, which cannot be described in detail here, contributed to the progressive loss of collective control (and declining rationality) of political and military decisions that took place in the last years of the regime. One cause of this development was the pseudo-religious popular belief in Hitler. The plebiscitarian foundation of Hitler's power, although it had been so highly effective for a long time, became a cause of the Führer's growing self-deception and self-adoration in the last phases of his career, as well as of the increasing arbitrariness of his decisions.

From the point of view of our subject, however, we have to be concerned not so much with the power structure of the Nazi regime, but rather with the extent to which there was accordance between the basic goals of the regime and the opinion of the people. With respect to this, I would like to discuss at least two significant examples: the Nazi war-making policy and the anti-Jewish measures.

Research into public opinion in the Third Reich which has been going on in West Germany in recent years now allows us to answer the questions raised more conclusively than would have been possible a decade ago. It is well known that the Nazi regime, as a substitute for the outlawed freedom of speech and press, laid extraordinary great stress upon permanent investigation of popular opinion, and demanded confidential reports on the mood of the people from many different agencies. The secret service of the SS developed from this a remarkable procedure by means of a system of confidants throughout the entire country. Not all by far, but many of the reports which came from this and other agencies have been preserved, and substantiate the following assertions.

Although war and conquest were at the core of Nazi ideology, here the will of the Nazi leadership coincided only to a very limited extent with that of the overwhelming majority of the German people. All the reports available to us confirm that the acute threat of war brought about by the Czechoslovakian crisis in 1938 by no means sparked off enthusiasm among the majority of the German population; on the contrary, more people reacted with great concern and fear. A year's Nazi propaganda certainly had stimulated an exaggerated national feeling of self-confidence, but it could not create the attitudes required by Hitler's ideological perception of the eternal struggle among nations for power and *Lebensraum*. Even with all the national *hubris* which the Nazis had brought about, the attitude of the majority of the population had little in common with such heroic projections. Most people affirmed and applauded the national policies of the regime as long as they were in line with the traditional policy of the revision of Versailles or with the idea of German hegemony in central Europe, and so far as they could be achieved through peaceful means and did not gravely threaten the restored social security and economic standard. Up till 1938, Hitler himself, despite all emphasis on national interests, had also stressed again and again the peacefulness of his intentions and his respect for the national interests of the neighbouring European countries. However deceitful this propaganda was, it left its mark on the German population. In a secret conference with representatives of the German press on 10 November 1938, shortly after the conference in Munich, Hitler himself complained that the year-long peace propaganda had brought the mentality of the population into an undesirable condition. Therefore, in future, the people would have to be prepared for more heavy national burdens and the endurance of war. Actually, the coming of the war in September 1939, and the Nazis' great triumphs in its first phase up till the summer of 1940, caused only temporarily stronger fears among the population, as the opinion reports indicate. Much more prevalent, especially during the campaign in France, was a new wave of national enthusiasm. But this reflected the fact that the German people had few burdens to carry in this first stage of the war; the success of the *Blitzkrieg* came quickly and brought about very few casualties on the German side. In consideration of the national mood, the government consciously practised restraint over war-related cutbacks in supplies; thus the consumer economy still flourished. War still seemed to be compatible with a peace-time economy. The Nazi regime also proceeded much more carefully than in the First World War with the military recruiting of workers and farmers, so as to avoid overburdening and discord among the people at home. In this phase war still appeared to large portions of the population to be a happy-go-lucky event and, under these conditions, produced additional patriotic enthusiasm for the regime. The decorated war heroes of the army, the air force, and the marines became adored models for the Hitler Youth, while the reputations of the local party representatives sank in their eyes. The patriotic emotions remained throughout the entire war, besides the *Führer-nimbus*, the strongest supports of integration. But from the beginning of the war against the Soviet Union in 1941, the foundations of integration changed considerably. More and more people began to ask what the expansion of the war to places so far away from Germany had to do with the national interests of their country. The canvassing among the rural population regarding future colonial settlements in the conquered eastern territories met mostly with reserved and quiet rejection, as did the attempts to encourage industrial enterprises to invest in or to establish new branches in the eastern territories. The Empire of a master race was for the majority of the people at best a lot of fancy talk, and was not followed in practice. The civil servants who were transferred to the east felt that this was more of a disciplinary transfer and showed little pioneering spirit. Despite all its propaganda, the regime had been unable to transend the perceived needs of the population for a civilized life; the regime's ability to mobilize support for its policies had clear-cut boundaries when it touched basic existential interests. And after Stalingrad, the government fell into a growing crisis of legitimation. The longing for peace became, as the opinion surveys show, the dominating element in the mood of the people. The popular approval which greeted Goebbels's proclamation of 'total war' in his demagogic speech at the Berlin *Sportpalast* on 18 February 1943, was only apparently in contradiction of this. What was applauded here was mainly the turning away from the plainly too optimistic official portrayal of the war, which in the months before had produced more and more criticism from the population. Also applauded was the announcement that no distinctions would be made any more between rich and poor as regards the war effort at home. Many people mistakenly believed that the total war effort would lead to a more equitable distribution of burdens and a more credible form of National Socialism. But this hope disappeared quickly. Besides the massive number of human losses, which were registered in almost every family, the increasingly harsher consequences of the Allied bombing attacks and the faster growing lack of provisions caused an increasing mood of fatigue and exhaustion. The war enthusiasm of 1940 had long since disappeared. There remained nevertheless a will to hold out to the end, partly apathetic and partly pertinacious. With many, this was motivated by the wishful belief that the sacrifices made in the war effort could not all be in vain, and the feeling that one had to prove oneself worthy through the further fulfilment of one's patriotic duty. For most people, this had nothing to do with the Nazi principles of indoctrination. And it was still rather important that the regime, almost till the end, was very concerned with keeping the material burdens and restrictions imposed on the civilian population as small as possible. For example, despite the proclamation of total war, women were more widely exempted from compulsory service for the arms industry in Germany than in Great Britain. At the expense of millions of foreign workers and forced labourers and the economic exploitation of the occupied areas, the regime could afford this relative sparing of its own population. Despite rising criticism of the party, which later also included criticism of Hitler himself, there was, however, no ensuing widescale support for active resistance or sabotage. Only in the very last weeks of the war, in the western and southern portions of Germany, were numerous local attempts made to stop the senseless fighting for the defence of cities and villages against the American or British troops. To explain this lack of resistance, it is certainly not sufficient to refer to the severe punishments for all forms of opposition, which were indeed massively strengthened during the second half of the war. The apathetic but continuing obedience till the very end was also determined by a frightened consciousness in the population of the catastrophic and criminal ends of the regime, in which nearly everyone had become entangled, either directly or indirectly. Most people knew nothing complete and nothing exact about these crimes, but they had a sufficiently substantiated suspicion or knowledge of them. This leads to our second question regarding the attitude of the German population towards the anti-Jewish policy of the Third Reich.

With the unification of Germany and the founding of the German Reich in 1871, the Jews in Germany had achieved full equality. The last major pogroms in the tradition of the medieval persecution of Jews took place in Germany in 1819, when famine and epidemics plagued the land. The Jews, as so often in previous centuries, were the scapegoats and had to pay for this. In 1933, that lay in the distant past, so it seemed. Since the beginning of the nineteenth century, the rule of law and principles of culture had appeared to be a strong safeguard for the Jews in Germany; this had been a major cause of the massive immigration of Jews into Germany, and their assimilation into German society. When the question of Germany's eastern borders arose at the peace negotiations in Versailles in 1919, German delegates could argue with conviction that the pogrom-boundary in Europe was identical with Germany's eastern border, and that shifting of it would thus not be advisable. Most of the Jews living in the German–Polish border area felt the same way. The majority of those who voted in the plebiscite in Upper Silesia chose Germany over Poland, because they considered Germany the land of greater opportunity and freedom, and greater culture and integrity. The constitution of the Weimar Republic, which gave Jews access even to the civil service, marked the definite completion of the emancipative process. But just at this point a new form of ideological anti-Semitism, already developed on a small scale prior to 1914, now made itself noticeable to a larger extent as a result of national and social depression. Anti-Semitism became an ideological element of the political Right. It had little basis in reality: the proportion of Jews in the population of the German Reich had been declining since the 1880s, and in 1925 the Jews constituted less than one per cent of the total population. The independent Jewish small businessmen were especially hard hit by the inflation and the economic crisis. The over-proportionate influence of Jews in trade and in some liberal academic professions, especially among doctors and lawyers, was certainly a fact, and was a basis for the already existing anti-Semitic social envy in the middle class and in the rising generation of non-Jewish academics. But the expansion of anti-Semitism during the Weimar era had only a little to do with this. In those cities and towns where Jews formed a large segment of the population, the relations between the Germans and the Jews were, even in the first years of the Nazi era, for the most part relatively good and hardly hostile. Characteristic of the new anti-Semitism was its abstract ideological form. It was not directed so much against individual Jews as it was against certain phenomena in social and cultural life which were derogatorily termed 'Jewish'. It was here that the national resentment towards the so-called rootless Jewish intelligentsia took hold. And it was applied to just those Jews who made considerable contributions to German culture, in the press, theatre, literature, and film. For this very reason, they were seen as downright devilish personifications of Jewish 'mimicry'. The *völkisch* anti-Semitism was directed not so much against the Jews as an exclusive religious or national group, but rather much more so against the assimilated wealthy Jews and the civilization-bearing classes. And it was especially full of hatred towards the élitist form and culture of German-Jewish intellectualism, as it had been formed in the urban cultural centres of Berlin or Frankfurt.

But the mass resonance of this *völkisch* anti-Semitism during the Weimar period should not be overestimated. It was more widespread among conservative intellectuals, nationally minded students, and secondary schoolmasters than in the broad masses of the people. It was at its weakest among farmers, although they were often in direct economic and social contact with Jewish cattle-, grain-, and wine-dealers. As a reaction to the leftist attempt at a revolutionary putsch in Munich in 1919, anti-Semitism in Munich and Bavaria in the early 1920s temporarily found mass support. Even so, the dominant element of the mass flocking to the Nazi party at the end of the Weimar Republic was not, as already mentioned, due to an equivalent amount of anti-Semitism. A large part of the German middle-class population sympathized with the fundamental idea of a reduction and repression of the Jewish influence in certain areas of public, professional, and economic life. Thus they had little objection to the first legal measures taken by the regime in the pursuit of this goal. But despite the tremendous production of anti-Jewish agitation, it was only to a small degree that the government succeeded in making its anti-Semitic racialist theories popular. Gestapo reports from the years 1936 and 1937 repeatedly criticized the rural population for lacking a racial consciousness. The farmers only broke their economic ties to Jewish traders when massive economic pressure was exerted on them. The party-sponsored boycotting of Jewish businesses, which had been staged at regular intervals since April 1933, had only little resonance among the population. And in 1935 when the Nuremberg laws forbade Jews any longer to hire or employ non-Jews as household servants, there was a remarkable flood of requests from such employees in Munich and other cities, asking for special permission to remain in the Jewish households. Popular support for the anti-Jewish policy of the Nazi regime came to a definite end with the staged pogrom of the *Reichskristallnacht*. The overwhelming majority of the German urban population, which witnessed the burning of synagogues and the looting and destruction of Jewish stores and businesses, reacted with utter shock and disbelief. Some public protests and relief actions for persecuted Jews ensued. But at the same time there was a fatal psychological constellation at work, suppressing wider public protest. From all the official manifestations of the Nazi regime, the demonstrations, the proclamations, and the propaganda, people knew that anti-Semitism represented the fanatically fixed party policy and that any signs of friendliness towards Jews would mean public exposure to condemnation, if not something worse. This intimidation by the radically anti-Semitic standards of the regime, together with the widespread elements of moderate anti-Semitism, caused the vast majority of the non-Jewish poulation to keep silent and simply let things happen. After the *Reichskristallnacht*, a series of further ordinances permanently reduced the Jews' civic rights even more. And finally, after the introduction of the Yellow Star in September 1941, the Jews who had not yet left Germany had been officially branded and made into outcasts. Giving one's help and support to Jews under such drastic conditions, which fortunately did occur quite often in Berlin and elsewhere, or denying such help at that time, was hardly a question of anti-Semitic or non-anti-Semitic attitude. It was simply a question of the amount of civil courage one was ready to summon up, to help these outlawed people despite threatening sanctions. For a large portion of the German population, the fact that they did not help the Jews and their lack of a critical consciousness for what was happening to the Jews was the limit of their involvement in the anti-Jewish policy of the Nazi regime. The ultimate step in the carrying out of this policy, the deportation and extermination of the

Jews, no longer had anything to do with the will of the overwhelming majority of the German population. And for this reason, it was carefully kept secret from them. The radical raising of anti-Semitism to a racialist theology, and, even more so, the organized murder of the Jewish people, were elements of Nazism in which the mass of the population did not take part. At the same time, however, they did little to prevent a development leading to this 'final solution'.

On the other hand, it is a fact that in the persecution of Jews, of political opponents, and so-called undesirable elements by the Nazis, the regime did not profit only from the silence and passivity of the people, but also from a considerable amount of voluntary co-operation. With all we know about the forms, contents, and effects of Nazi education and schooling, we can say that this active following was caused probably less through the indoctrination by the Nazi *Weltanschauung* than by the systematic undermining of Christian and humanitarian foundations of morals and conscience, achieved through propaganda and agitation praising strength, courage, and steadfastness, along with other heroic male virtues, and denouncing sympathy for fellow men as a weakness and an over-exaggerated feeling for humanity. That was the other side of the effective mobilization of the national mass-society, which we otherwise characterized under the aspect of social modernization. The Nazi *Weltanschauung*, totally empty of constructive ideas for a new order of society, was the more effective in bringing about the negation of traditional values, the questioning and undermining of traditional social norms, and the breakdown of ethical and cultural values that could have been resistant to the ideology of efficiency and purpose, which had become prevalent in the aroused nationally egoistic mass society of the Third Reich. German civic and political culture even before the Third Reich was characterized by a lack of a sufficiently long and a sufficiently well-developed tradition of humanism and enlightenment before the rapid formation of an industrial society, with its specific and dominant norms for effectiveness and discipline, that was linked with the mainstream of a German national ideology that was anti-Western. This civic and political culture suffered a further catastrophic breakdown of its already weak humanistic foundation during the Nazi period. This is the most fatal aspect of the mobilization of the people which Nazism promoted. One should not forget, however, the coercive forces of the totalitarian dictatorship and, especially during the war, the intensified repression of all oppositional opinions. For many of the German people, particularly of those 56 per cent who did not vote for Hitler in 1933, the

Nazi reign was also a kind of foreign domination. A part of this 'other' Germany had demonstrated their rejection of the Nazi regime even before the war through courageous actions. At least 150,000 German Communists and Social Democrats were put into concentration camps before 1939. About 40,000 Germans emigrated for political reasons in the first years of the Third Reich. (This does not take into consideration the much larger number of emigrants from racial reasons.) Between 1933 and 1939, approximately 12,000 persons were convicted of high treason, for their efforts to overthrow the government. And in the same period, about 40,000 Germans were sent to prison for lesser political crimes. These partial statistics clearly show not only the extent of the terror exercised by the Gestapo and the penal courts, but also that the Nazi regime was by no means simply tolerated by the German population without any resistance. During the war, penal sanctions became considerably stricter on all levels. The number of offences entailing death-sentences in the penal code rose from three in 1933 to no fewer than forty-six by the end of the war. And the German civil courts alone, leaving aside the military courts, pronounced 15,000 death sentences during the war. Since 1942, there had been thousands of death sentences simply for public criticism of the regime, especially if it could be proven that the accused had a hostile attitude towards Nazism. Before Adam von Trott became a victim of the brutal executions which resulted from the unsuccessful assassination attempt against Hitler on 20 July 1944, other smaller or larger conspiratory groups had been discovered, and their members shamelessly executed. One example is the *Rote Kapelle*, a large group with national-communist goals, uncovered in 1942, another the Munich student group, *Die Weiße Rose*, which up till the beginning of 1943 had distributed pamphlets which indicated very clearly the criminal character of the Nazi regime.

Between the minority of convinced fanatical followers of the Nazi regime on the one hand, and the active opponents of the regime on the other, stood the majority of the German people. This vast majority till the end felt conflicting emotions in the presence of a regime which could use propaganda with such dexterity in evoking positive feelings towards the nation or resentment towards a particular group. Between partisanship and resistance there was a wide range of different attitudes. There was the partial opposition of Christians and clergymen, when it concerned their religious conviction and the church. There was the resistance of writers and artists who retreated into 'inner emigration' or into non-political work in order to preserve their intellectual immunity.

There was also the resistance of public officials and judges, who in many cases gave legality and loyalty to their profession priority over humble and submissive conformity. For the most part the immediate interests of those involved in opposition, together with their limited competence and insight, meant that those many little individual acts of opposition were not elevated to the level of fundamental resistance. And the compulsory co-ordination of almost all publicly effective social institutions resulted in a splitting of collective resistance. The isolation of the individual brought about by the totalitarian structure of the political system led to the dismantling of the potential for resistance, and robbed it of its social and political relevance. Nor did the break between the regime and most of the population, which began to develop after Stalingrad, lead to an active opposition; it led rather to a retreat into the private sphere. After a decade of having been permanently mobilized for the regime, society rediscovered the importance of the individual and private realm, as its material values were being smashed to pieces in the latter part of the war. Thus began the internal withdrawal of the majority of the people from the Third Reich and their development of the norms and standards of post-war Germany society. The mobilization of the German people under Nazism and the immediate consequences of the war did away with many of the old structures of German society: the traditional forces as well as the credibility of the old social standards and norms. This was a new opportunity and, at the same time, a heavy new burden.

6

THE THIRD REICH AND THE GERMAN LEFT: PERSECUTION AND RESISTANCE

TIM MASON

JULIUS LEBER, Social Democrat, was arrested by the Nazi regime for the third and last time on 5 July 1944, fifteen days *before* his friend and fellow conspirator Stauffenberg made the unsuccessful attempt upon Hitler's life. Leber held out under torture for these two weeks; he did not know precisely how many days would pass before Hitler might be assassinated and he himself liberated from his persecutors. He did know that such an attempt was very near. Leber's arrest made it the more urgent for the conspirators to act against Hitler, and his resilience left them a chance. In these intervening days Stauffenberg repeatedly said that Leber must be got out, that the coup needed his political leadership.

Leber's successful defiance of his interrogators may have been facilitated by the improbable and misleading circumstances of his arrest. This resulted not from his long-standing collaboration with Goerdeler, Beck, Stauffenberg, and the Kreisau Circle, about which the Gestapo then knew little or nothing, but from his first meeting with the central leadership of the underground German Communist Party. Since the German invasion of Russia in June 1941 the Gestapo had already destroyed two such Communist organizations, but during 1943 Anton Saefkow and Franz Jacob, both party veterans who had served time in prison and concentration camp, built up a new secret national organization, the main strength of which was based in the capital city, the north-coast ports, and in Saxony.

Leber had never been friendly towards the Communist Party. However, he and a few other members of the military conspiracy thought it necessary to form at least an impression of the strengths and intentions of this Communist resistance before they launched their own *coup d'État*. No such approach had been made before. Many of Leber's collaborators advised against it, either because they did not want to have anything to do with Communists, or because they considered such contacts to be a bad security risk. The latter were proved right. At the first meeting on 22 June 1944 Leber's own concern about security was aroused by two incidents: Saefkow and Jacob were accompanied, unannounced, by a third Communist; and, contrary to prior agreements, Leber was identified by name during the discussions. He was also puzzled by the alacrity with which the Communists accepted liberal points in his programme. Anyway, he did not attend a second meeting on 4 July, but his prudence came too late. Like many earlier Communist organizations, that of Saefkow and Jacob had been infiltrated by a Gestapo agent (in fact the third man in their delegation). Everyone at the second meeting was arrested, Leber was picked up the next day, then over 100 Communists, many of whom were later executed. The meetings may at first have looked like yet another attempt by the two parties of the Left to resolve the conflicts between them, conflicts which had enfeebled the German labour movement since 1918. With the arrest after 20 July of Leber's fellow conspirators, who came from all points of the political spectrum, it became clear to the Nazi authorities that something larger than this had been at stake. Whatever this might have been, it ended, like so many different resistance efforts before it, in persecution, death, and destruction.[1]

The story of the Third Reich and the German Left is that of a bitter political war. Since this war turned out to be more one-sided than any of the participants expected, it is in the first instance the story of an unexampled political persecution. The Nazi regime set out to obliterate the German Left. The Social Democratic and Communist Parties were outlawed, the trade unions destroyed, and the many smaller socialist groupings were hunted down in the same manner; newspapers and journals were closed, writers of the Left imprisoned or forced into exile and their works publicly burnt. The labour movement's wide range of leisure and cultural associations were either dissolved or reconstituted under close Nazi control: from the burial clubs of the Free Thought League to the Workers' Chess Groups, the internationalist Esperanto clubs of the Left and the federation of workers' hiking groups (the Friends of Nature), every form of organization which could serve as a legal focus for socialist or communist activities of any kind was either outlawed or taken over by the Nazi regime. Not only were the offices and printing presses of parties and unions sequestrated; the vast properties of the consumer co-operative organization and of the labour movement's own insurance, housing, and welfare concerns were placed under new management, and the extensive facilities of the workers'

[1] See Peter Hoffman, *The History of the German Resistance 1933–1945* (London, 1977), 362 ff, 377 f.

sports associations and theatre movement were put at the disposal of Nazi organizations.[2] Deprived of all material resources which they could call their own, Social Democrats and Communists were denied any place where they could meet and concert: the physical spaces left to them were private homes, woods and fields, corners in public places like railway stations, parks, and dole-queues, and, as employment recovered through the 1930s, the workplace.

The men and women who used such spaces to keep the causes of Social Democracy and Communism alive were hunted, and from the first months of the Third Reich they knew that they would be hunted. The police used sophisticated techniques of detection and repression which were decisively reinforced by denunciations from among the populace at large. All forms of behaviour which looked eccentric or suspicious were liable to be denounced. In the first two years these controls were supplemented by crude and violent dragnet operations in which the police and SA sealed off working-class quarters of towns and combed them house by house. Within months of Hitler's installation as Chancellor the penal code had been rewritten in such terms as to criminalize every conceivable left-wing political activity, either as malicious propaganda, or as the illegal refounding of a political party, or, most frequently, as preparation to commit treason. The police could hold people on suspicion. Investigatory arrest on political charges were almost routinely accompanied by beatings, often by torture to elicit the names of accomplices, and it normally lasted a long time. In the 1930s the subsequent court proceedings were normally followed by long terms in prison, often with hard labour; in the war years death-sentences were normal. If the courts were not severe enough the Gestapo took resisters into the concentration camps. The persecution of the Left during both the first and the last months of the Third Reich was so intense and so arbitrary that no comprehensive census of the victims can be made. Those arrested probably numbered nearer to 200,000 than to 100,000. It was numerically the largest sector of the German resistance. Few militants escaped arrest, and among the tens of thousands who were done to death were many who had left Germany to continue the struggle from neighbouring countries in the 1930s, only to be picked up by the

[2] The great scale and variety of this working-class associational life and its importance for the conduct of politics are well brought out by local studies: Gerhard Hetzer, 'Die Industriestadt Augsburg. Eine Sozialgeschichte der Arbeiteropposition', in Martin Broszat *et al.* (eds.), *Bayern in der NS-Zeit* iii, (Munich and Vienna, 1981) 182–7; and Klaus Tenfelde, 'Proletarische Provinz. Radikalisierung und Widerstand in Penzberg/Oberbayern 1900–1945', ibid. iv (Munich and Vienna, 1981), 154–60, 225–7.

Gestapo or by the police-forces of collaborationist regimes as the Third Reich established its power across the whole of continental Europe. This enterprise of persecution and repression was conducted on a huge scale and for the most part in a remorselessly methodical manner. It was maintained throughout the years 1937–41 when the major underground resistance groups appeared to have been smashed and the exiled parties of the left were in increasing disarray, and it was accompanied by a constant vilification of the political ideas of Socialism and Communism and of the role that these movements had played in German politics in the preceding decades.

Why did the Nazi regime set out to annihilate the left, rather than merely subjugating it? A full explanation for the politics of visceral hatred is always hard to construct, but the conscious motives of leading Nazis for pursuing a strategy of annihilation were both strong and clearly articulated. Hitler, together with most of the nationalist right, believed that the German armies had been stabbed in the back in 1918, that victory had been snatched from them by socialist revolutionaries and strike leaders who had caused the home front to collapse. The lesson drawn from this powerful political myth was both simple and radical: all such 'subversive elements' had to be eradicated *before* Germany's next bid for imperial dominance. That was one essential condition of military success. During the Second World War Hitler, Himmler, and Göring repeatedly emphasized that there would be no second 1918 because the traitorous internationalists of the left had been obliterated. (There were indeed no risings in 1945, but it is by no means clear that this was the main reason.)

This motive for extreme persecutory rigour was reinforced by a second. Nazism saw politics in terms of élites and masses, leadership was everything. These axioms were projected on to the opposition, with the result that the parties and movements of the left were perceived as consisting of small cadres of ideologically committed, skilled, and fanatical leaders, and a large mass of basically gullible supporters. Thus, if the so-called ideologues, scribes, and wire-pullers could be simply removed and their organizations destroyed, the rank and file of the labour movement would soon be persuaded that it had been misled in the past, that its interests were those of the German people as a whole, that it should transfer its loyalties to the Nazi leadership. The belief that the mass of the population, including the working class, was fickle and superficial in its political allegiances, that its outlook could be remoulded by dictatorial propaganda, was central to Nazi politics.

Precisely because people were supposedly to be easily led, it followed that opposing leadership groups had to be not merely broken, but wiped out in a campaign of perpetual repressive vigilance. The terror was pre-emptive: the Left had to be denied the *possibility* of influencing people. Hence the stream of death-penalties for minor expressions of opposition after 1941.

The war against the Left was also fuelled by more mundane political calculations. First, it was popular. In the spring of 1933 it was very popular in the Nazi movement itself. It was the local formations of the party and the storm-troopers which dealt the heaviest blows to the working-class parties and the trade unions with the wave of terror which they unleashed in the weeks following the Reichstag fire. Offices were ransacked, labour leaders beaten, publicly humiliated, imprisoned, tortured, and killed in a prolonged outburst of brutal fury; now unimpeded, often indeed assisted, by the police, the Nazis settled accounts with their erstwhile enemies of election-campaigns and street-battles. In this crude form the popularity of the cause brought with it the passing risk of civil war. The popularity among non-Nazis of more orderly forms of anti-Marxism was, however, pure gain to the regime: nothing did more to align conservative forces and the Roman Catholic Church with the regime than its rigorous anti-Communist and anti-Socialist policies. These forces of the traditional right remained bound to the Nazi leadership by their anti-Communism long after the regime had begun openly to attack their own beliefs and positions of power and influence. The sense of having a common enemy on the left was decisive in holding the different élite groups of the Third Reich together.

Then, because it was popular, the assault on the Left could also serve as an effective pretext for other political operations by the dictatorship. It was behind this façade, for example, that the federal structure of the constitution was abolished and a centralized state machine constructed in 1933. And in the later 1930s Himmler and Heydrich deployed the 'threat from the Left' in order to acquire for the SS full extra-legal powers as the morals police over German society and let loose a reign of terror against drunkards, homosexuals, the work-shy, and 'deviants' of all kinds.[3]

Thus the war against the Left was an integral part of Nazi rule. It mobilized many different fears and hatreds, served many different economic interests and political purposes. The categorical negation of Marxism gave a spurious substance and identity to Nazi ideology, and

[3] See Detlev Peukert, *Volksgenossen und Gemeinschaftsfremde* (Cologne, 1982), 11–12.

the invocation of Bolshevism as a terrifying Jewish pestilence legitimized brutal measures of political hygiene. To a considerable degree the power of the regime was parasitic upon the supposed threat from the enemies which it persecuted so implacably.

In the summer of 1933 the first response of the Communist movement and of parts of Social Democracy was to rescue their organizations into illegality and from there to wage an *offensive* political campaign, regrouping and mobilizing their supporters to denounce the new regime. The main focus and instrument of this form of mass resistance was the underground press. Few preparations had been made, but both inside Germany and beyond the frontiers the large-scale production of newspapers, journals, pamphlets, and leaflets was swiftly achieved, not least thanks to the ingenuity and professional skills of printing workers. The distribution of this literature then became the principal activity of the clandestine party groups and organizations. The underground press served many different political purposes. Of necessity, it was now the main link between the cadres and the supporters, informing the latter of their party's diagnoses and strategies, encouraging them to believe that the dictatorship could soon be overthrown, suggesting arguments for voting 'no' in Nazi plebiscites and factory council elections, etc. (However, for obvious reasons, the tactics of the underground struggle could not be discussed or evaluated in the press.) Second, the illegal newspapers met the increasingly vital need of the rank and file for information which they trusted: the suppressions and distortions of the Nazified media sharpened this hunger for facts and truths—the truth about the burning of the Reichstag, facts about the fate of leaders who had disappeared, about strikes which occurred despite their prohibition, or about the numerous corruption scandals in the Nazi movement. Beyond this the underground press documented the continued defiant existence of the Social Democratic and Communist parties: newspapers and pamphlets became symbols of loyalty, with individual copies passed from hand to hand—both emotionally and organizationally they were a substitute for the outlawed party. This specific form of mass resistance demonstrated the dependence of both parties upon the (basically democratic) techniques of mass communication and persuasion for their political influence; the Communist Party even hoped that its underground publications could win it new supporters. For the German labour movement, mass persuasion had been quite simply the essence of politics—it was what they believed in and knew how to do. Within two years, at the latest by the middle of 1935, it had become clear that this kind of politics could not be carried out under the conditions of Nazi persecution.

Some details should be given of this confrontation. During 1933, for example, the Communist party built up a new 'technical apparatus' for the production and distribution of literature. By the end of the year the Ruhr regional section alone had three illegal print-shops and half a dozen mechanical duplicators in operation. They produced copies of the national paper (*Die Rote Fahne*), a regional paper, and local broadsheets; over 10,000 copies of Dimitrov's defence speech in the Reichstag fire trial were printed as a pamphlet entitled 'Electrical Heating in the Home'. When the Gestapo smashed this organization in April 1934 it was immediately reconstructed: *Die Rote Fahne* continued to appear every ten days, and 2,500 copies were printed in Cologne, 12,000 in a Solingen print-shop (this latter turned out no less than 300,000 publications in a few months before the Gestapo destroyed it in November 1934). From the start of 1935 the Communists party moved bulk printing over the Dutch border, and newspapers were smuggled into Germany in quantities larger than the underground cells could cope with. However, there was a police agent among the smugglers and this whole network was broken up in April 1935. Therewith, the Ruhr regional organizations of the Communist Party, which had once numbered well over 25,000 members, was reduced to eleven small groups, and they refused to continue with the same kind of resistance activity. In all fully one-third of the pre-1933 membership participated actively in the early resistance struggle and over half of these people were arrested (some of them twice), mostly on charges connected with the distribution of illegal literature.[4]

The Communist effort in this first phase was that of a bureaucratically centralized party aiming at the revolutionary overthrow of Nazi rule. The Social Democratic commitment to revolution, as expressed in the Prague Manifesto of January 1934, was less whole-hearted, in part a tactical attempt to bridge the deep policy differences within the party which had all but fragmented it in the first months of Nazi rule. In consequence of these disputes it was not so much the reconstructed remains of the party machine which became the main bearer of the resistance struggle from the summer of 1933, but rather a wide variety of local groups which often drew their strength from the party's ancillary

[4] Id., *Die KPD im Widerstand* (Wuppertal, 1980), part II, ch. 2.

organizations: the Socialist Workers' Youth, the Iron Front, the Republican Defence Force (*Reichsbanner*). The Social Democratic underground was less centralized and less uniform than that of the Communist party, but its characteristic activities were very similar: they focused around broadsheets, newspapers, and journals. The Red Vanguard group, for example, was formed very quickly in Berlin by a group of left-wing Social Democratic students and young people; they inspired similar groups in other cities, and collaborated with underground Liberal, Jewish, and Catholic circles. They ran off their weekly newsletter in several different places, soon, it is claimed, in a print-run of 20,000 in all; some were mailed to supporters. A postal error put the police on their track, and the organization was destroyed in November 1933. After serving ten years' hard labour, the leader, Rudolf Küstermeier, was put in a concentration camp.[5]

The principal organ of the Social Democratic underground, *Sozialistische Aktion*, was printed in Czechoslovakia and smuggled into Germany. In Hamborn (Ruhr), a woman school-teacher heard that an ex-party member had bought a large bakery, and she hit upon the idea of using the breadrounds to distribute *Sozialistische Aktion*. This operation was organized from June 1934 on by a local militant, who could already look back on a successful resistance career; the bakery-owner seems to have gone along with it as much because party men were diligent salesmen who could attract loyal customers as out of personal political conviction. Members of the underground transport workers' union brought *Sozialistische Aktion* into Germany in crates disguised as canned food or cigarettes, and 'Social Democratic bread' was soon being sold door-to-door in the mining communities of sixteen neighbouring towns. The owner of the bakery sacked a couple of Nazis in order to hire more Social Democrats; this was noticed, and the local Nazi party leader alerted the police. Gestapo headquarters was already familiar with similar groups of Social Democratic insurance salesmen in other parts of Germany; thus, a few double agents and some telephone taps led in June 1935 to several hundred arrests in the Hamborn area. Four of those arrested died under torture; 170 were put on trial.[6]

In Berlin there was a large concentration of lesser party functionaries and employees, and here, exceptionally, parts of the old party machine

[5] Günther Weisenborn, *Der lautlose Aufstand*, (Hamburg, 1962), 148 ff.; Annedore Leber, *Das Gewissen entscheidet* (Berlin and Frankfurt am Main, 1957), 33 f., 47 f.

[6] Kuno Bludau, *Gestapo geheim! Widerstand und Verfolgung in Duisburg 1933–1945* (Bonn-Bad Godesberg, 1973), 35–41.

maintained an illegal existence and distrubuted *Sozialistische Aktion* (fully 1,000 copies of each number during 1934). Successive waves of arrests during 1935 put an end to this large organization; thereafter those Social Democrats who were still at liberty in Berlin concentrated—quite effectively—on maintaining informal contacts with one another.[7]

Mass resistance built around the printed word had three major weaknesses as a stragtegy. First, the publications themselves were hard to disguise and constituted irrefutable proof of 'criminal' behaviour by those caught in possession of them. Second, their distribution required large organizations, the whole of which became vulnerable once the police had a single good lead; the bureaucratic structure of the Communist Party was quite fatal in this respect. Last, on account of its huge human cost, that is the early loss of many of the most energetic activists, mass-resistance depended upon a swift success or break-through of some kind. Hitler's consolidation of the regime through the murder of the SA leadership in June 1934 removed any chance of such a success and denied the Left a last opportunity for mass public action. What were the alternative strategies?

Walter Loewenheim, pseudonym Miles, founder of the independent socialist organization 'Begin Anew!', had clearly foreseen all these three problems before 1933. His solution, both to the conflict between the Social Democratic Party and the Communist Party and to the (correctly anticipated) stability of the Nazi regime, was to build up a new élite cadre organization on the most strict conspiratorial principles. He dismissed the Communist Party as a 'decimated apparatus for the illegal distribution of pamphlets'. In the interests of security he ruled out all such resistance activities, until a crisis of the regime should furnish the moment to strike. Where it was applied rigorously this conspiratorial strategy achieved a considerable negative success, i.e. the groups had a long life. Through its decentralization and its clandestine procedures, for example, the Berlin centre of 'Begin Anew!' survived a series of arrests in 1935–6, and only fell victim to the Gestapo at the end of 1938 when Fritz Erler and Kurt Schmidt broke their own rules by associating themselves with another dissident socialist group, the German Popular Front, which circulated 100 copies of a compendious political programme.[8] To take another

[7] Frank Moraw, *Die Parole der 'Einheit' und die Sozialdemokratie* (Bonn-Bad Godesberg 1973), 32–8.

[8] Kurt Klotzbach (ed.), *Drei Schriften aus dem Exil* (Berlin and Bonn-Bad Godesberg, 1974); H. Hellmann's obituary of Miles in *Internationale wissenschaftliche Korrespondenz zur Geschichte der deutschen Arbeiterbewegung* (1977, no. 2); Moraw, *Parole 'Einheit'*, 48–50.

example, Josef Wager, a skilled engineering worker, and Hermann Frieb, a tax accountant, were able to keep the Augsburg/Munich organization of 'Begin Anew!' in being for eight full years. They were preparing an armed uprising for the event of the—as it seemed, imminent—military defeat of Nazi Germany, when their links with the Austrian socialist underground enabled the Gestapo to 'roll up' their group in 1942.[9]

The difficulty with conspiratorial *attentisme* was that for many who chose this path the waiting became too long. Miles himself was among the first to decide that the right opportunity to act would not come until after the defeat of the Third Reich in war; thus his sense of responsibility for the lives of the resistance fighters then led him to try to liquidate his group in the autumn of 1935. Other groups which placed clandestinity before action found that members drifted away after successful, undetected underground careers of several years. Wager and Frieb lost five collaborators in this way at the end of 1935, and, in the same manner, the clandestine activities of the large Socialist Workers' Party network in Baden—which included regular secret meetings for political discussion, the internal circulation of a newsletter etc.—began to decline in intensity in 1936, two years before the organization was betrayed to the Gestapo.[10] The combination of ever-present acute danger with the conspiratorial imperative to remain silent, inactive towards the outside world, tended to sap energies and confidence. The rules of conspiracy permitted only one form of initiative, the recruitment of new members, but success in this respect could itself be a major risk, for the larger a network became the more likely was a breach of security; the brilliantly constructed Socialist Front in Hanover finally attracted the attention of the police in 1936 when it expanded into neighbouring towns. There were 1,000 arrests, 231 convictions.[11]

At the end of 1936 the exiled leadership of the Social Democratic Party admitted that its underground in Germany was 'in no way a decisive factor' any more.[12] The Communist Party did not practise self-criticism of this kind, but its abandonment of mass resistance and its turn to a

[9] Hetzer, 'Augsburg', 191–205; Heike Bretschneider, *Der Widerstand gegen den Nationalsozialismus in München 1933 bis 1945* (Munich, 1968), 107–21.

[10] Hetzer, 'Augsburg', 199 f.; J. Schadt, (ed.), *Verfolgung und Widerstand unter dem Nationalsozialismus in Baden* (Stuttgart, 1976), 92 f., 195, 199, 202–18.

[11] Erich Matthias (ed.), 'Der Untergang der Sozialdemokratie 1933', *Vierteljahrshefte für Zeitgeschichte* (1956, no. 2), 201 ff., 218–26.

[12] Moraw, *Parole 'Einheit'*, 24.

Popular Front strategy late in 1935 pointed towards a similar assessment of the balance of forces in Nazi Germany. Serious though it was, the problem of the numbers of people still available for resistance work came to be overshadowed by the question of what those sorts of people who were Social Democrats or Communists could actually *do* to damage the hated regime. A police spy who reported on the regular informal conversations of a group of Düsseldorf Communists for several years from 1938 recorded not a loss of conviction but a sense of impotence cast in the form of speculations about outside interventions which might bring Nazi rule down.[13] The political initiative seemed to have passed irretrievably to the dictatorship and, after 1938, to the governments and armed forces of the other major powers. The Social Democrats in exile did in fact reach this bitter conclusion. For resisters in Germany 'doing something' in this situation meant more than the anyway dangerous and demanding run of normal resistance activities, such as it was by then: maintaining informal contacts, giving support to the families of those in gaol or concentration camp, attending the funerals of dead comrades, listening to Radio Moscow or the BBC and passing the news on, or even, for the few, getting political reports or party or union officials or victims of persecution across frontiers into safety. To damage the regime new forms of offensive direct action had to be invented, forms which should not be suicidal in inviting certain detection and retribution. This proved to be an almost impossible task.

Counter-violence, armed attacks on agents of the regime, or desperate acts of insurrecton did not seem to offer any kind of solution to the problem of what to do. It is very difficult to analyse the issues involved in this question. Apart from the instinct for self-preservation, three types of reason seem to have told against resort to violent resistance. First, the regime always took vengeance for such acts in an indiscriminate manner. The Gestapo responded, for example, to the killing of one of their spies with the murder in February 1934 of John Schehr and three other imprisioned leaders of the Communist Party; this barbaric act seems to have been intended as a terrible deterrent warning to the party's underground groups against adopting violent methods. Similarly, in May 1942 a Communist resistance organization with several young Jewish members set fire to the anti-Soviet exhibition in Berlin: not only were the activists detected and executed, but the Gestapo rounded up 500 Berlin Jews at random, shot half of them at once, and sent the rest to their

[13] Peukert, *KPD im Widerstand*, 330–2.

deaths in Sachsenhausen concentration camp.[14] Thus, through its control of the media the regime could deprive acts of violence of any demonstration effect, and then make innocent people pay the price by escalating the repression with reprisals. This strong moral and political restraint upon violent resistance was reinforced, second, by the political culture of the German labour movement. In almost all its different forms the German labour movement had set very high store by organizational discipline and self-discipline, by strategies and tactics of collective action which carefully selected the means appropriate to specific ends, and were then strictly held to. The trade unions in particular had dinned these values into their own more volatile members since the 1890s (and this is one reason why there were no large insurgent strikes in the Third Reich). By this canon spontaneous rage and expressive violence were politically immature and dangerous to the movement.[15] Thus it was not the political underground of the left, but street gangs of young proletarians with no experience of the disciplines of the labour movement, who did try during 1944 to wage a sort of guerrilla warfare against the Nazi regime: in the Cologne area alone they killed 18 officials and soldiers, including the city's Gestapo chief, and 11 other persons, before some of their leaders were publicly executed. However, their aggressively youthful ethos, which showed scant regard for the traditional decencies of the labour movement, the immediacy of their perspectives, and the fact that they constituted an underworld living partly by theft and collaborating with army deserters and escaped foreign labourers, all cut them off from the old left.[16] Then, third, beyond this distrust of spontaneous anger, was the moral fact that murder, arson, and the like did not come easily to people who had learned their politics in either the Social Democratic or the Communist Party. It is a debatable point, but I do not believe that the bulk of German Communists differed from Social Democrats in this respect after 1933.[17] Commitment to a basic code of decency was one of the reasons why they were opponents of Nazism. This vital point can be demonstrated in many different ways, perhaps none more illuminating

[14] Horst Duhnke, *Die KPD von 1933 bis 1945* (Cologne, 1971), 479 f. Hetzer, 'Augsburg', 160–3, gives many examples of violent and indiscriminate reprisals by the police following what were presumed to be acts of violent resistance by Communists in 1933–4.

[15] This theme is strongly emphasised by Barrington Moore, Jr., *Injustice. The Social Bases of Obedience and Revolt* (London, 1978), part II, esp. 219 f.

[16] Arno Klönne, 'Jugendprotest und Jugendopposition', in Broszat *et al.* (eds.), iv, 584–613; Detlev Peukert, *Die Edelweißpiraten* (Cologne, 1980).

[17] There were some differences before 1933: see Eve Rosenhaft, *Beating the Fascists? The German Communists and Political Violence* (Cambridge, 1983).

than the actions of a group of Social Democrats and Communists who seized control of the small mining-town of Penzberg in Bavaria on 28 April 1945. They believed, wrongly, that Munich had just surrendered and they wanted to save the mine from destruction. They spared the lives of the Nazi mayor and the senior police officer; after their rising had been broken by a passing army unit which shot the ring-leaders, it was these two men who helped to organize a massacre of politically suspect people in the town on the last day of the war. The rising did not fail *because* the miners' leaders were insufficiently ruthless; and if they had killed the mayor and the police officer the Nazi retribution might have been even more terrible. They did what they thought was necessary and right, acting with determination, civic responsibility, and decency. For people like this terroristic resistance and revenge had never been an option. (This is part of the reason why most anti-Nazis did not settle accounts with their persecutors in Germany after May 1945.)[18]

Were there any other options? Industrial sabotage could hamper the Nazi war effort, and it was not easy to detect. After 1941 Communist resistance groups, which were by then mostly organized around the work-place, gave high priority to damaging plant and raw materials and to turning out faulty products; the Uhrig organization in Berlin issued detailed instructions along these lines.[19] The effectiveness of industrial sabotage cannot be properly assessed, but it seems unlikely that such tactics had a wide appeal among those workers who did not accept the regime: one can suggest as a hypothesis that they ran counter to elementary patriotic sentiments in a very specific way, for the immediate price of successful sabotage was paid not by the regime's leaders, but by common soldiers, by workers in uniform. Pride in one's work was probably also a basic inhibition.

In order to attack the Nazi leadership directly, it was almost essential for resistance groups to have positions of power, or at least cover, within the regime itself. Here the organizations of the Left were at a major disadvantage. It was, however, only *almost* essential, because there was

[18] Broszat *et al.* (eds.), i (Munich and Vienna 1977), 322–5; Tenfelde, 'Penzberg', 376–81. Cf. the manner in which the prisoners in Buchenwald treated their guards after they seized control of the camp on 11 Apr. 1945, as recounted by one of their leaders, Walter Bartel, in Christoph Kleßmann and Falk Pingel (eds.), *Gegner des Nationalsozialismus* (Frankfurt am Main/New York, 1980), 243–51. I have tried to explore this question of moral codes more fully in a contribution to the *Festschrift* for James Joll, ed. Anthony Polonsky *et al.*, *Ideas into Politics* (London, 1984).

[19] Weisenborn, *Aufstand*, 158–60; Bludau, *Gestapo geheim!*, 158–64. Bludau also gives two examples of seemingly unsuccessful demands being made of communist sympathizers that they commit acts of industrial sabotage as a systematic tactic, 151–5, 165.

one remarkable exception: after months of solitary and precise preparation, Georg Elser, a cabinet-maker, almost succeeded in assassinating Hitler with the bomb which he built into a pillar of the Bürgerbräu beer-cellar in Munich in November 1939. Elser had no active political affiliations. He decided in September 1938 that Hitler meant war: war was a terrible crime, and this gave Elser a justification, indeed a duty to kill him. Elser carefully calculated his opportunity from Hitler's annual speech in commemoration of the 1923 putsch, and his lone effort showed that serious attempts at assassination could be made (at least until November 1939) from below, from quite outside the power structure itself.[20] With the exception of Beppo Römer, whose National Bolshevik group made some obscure assassination plans in 1941, the highly trained professional revolutionaries of the Communist underground did not make such attempts. Why not, remains in some ways a puzzling question.[21]

The one Communist group which did have members near to the centres of power of the regime was the so-called Red Orchestra. The two leaders had unusual biographies. Harro Schulze-Boysen had been a national revolutionary student leader before 1933 and was persecuted for a time by the Nazis; in the later 1930s, after gaining a post with officer rank in the Air Ministry, he developed strong sympathies for communism and the Soviet Union. Arvid Harnack, his chief collaborator, is a less transparent figure: he was an economist, and, despite the fact that he was an intellectual Marxist with a lively interest in Russian economic planning, he advanced during the 1930s as a civil servant in the Ministry of Economics; he joined the Nazi party as a cover, and it is conceivable that he was passing information to Soviet agencies in Berlin throughout the decade. The two men's strong family connections with the old German élite both helped them to make careers under Nazi rule and marked them off from the vast majority of Communists (more precisely, from those who had not gone into exile). The milieu of the Red Orchestra was that of a radical *bohème*—artists, students, intellectuals, and Communist manual workers, men and women united by their free life-style and by a strong antipathy to Nazism. The latter they expressed both before and after June 1941 in a variety of conventional

[20] Anton Hoch and Lothar Gruchmann, *Georg Elser: Der Attentäter aus dem Volke* (Frankfurt am Main, 1980); J. P. Stern, *Hitler. The Führer and the People* (London, 1975), 138–53.

[21] On Römer, see Hoffmann, *Resistance*, 30 ff. One aspect of the problems of assassination is illuminated by Hoffmann's subsequent study, *Hitler's Personal Security* (London, 1979).

resistance activities, especially the political organization of foreign workers. On the German invasion of Russia, however, Schulze-Boysen and Harnack decided that the best contribution they could make to the overthrow of Nazi rule was to pass military and economic intelligence to the Russian secret service. For twelve months they showed great inventiveness and daring in the use of portable radio transmitters, and then German counter-intelligence caught up with them; many were executed. Their treason was probably not of much practical use to the Red Army (though their reports might have become more accurate and comprehensive in time). Some German soldiers certainly lost their lives as a result of the Red Orchestra's actions, but they did not affect the outcomes of major battles. This form of resistance was not acceptable to all members of the group, and since 1945 it has excited bitter controversy, because, unlike the contacts of conservative resisters with the Western powers, the treason of the Red Orchestra was not part of a plausible larger plan to save the peace or to end the war. Still, people did what they could, and this clearly was one form of resistance, for the members of the Red Orchestra acted only out of political conviction.[22]

The last set of decisions about what to do brings us back to Julius Leber and the other Social Democrats who joined the conspiracy behind the attempted coup of 20 July 1944. By the late 1930s these men had come to the conclusion that they themselves, the labour movement on its own, could not offer effective resistance to the Nazi regime, and could certainly not overthrow it. Their answer was to construct alliances with the political representatives of that organization which, as they saw, could alone do something: the armed forces. At the start of their collaboration, Wilhelm Leuschner, the trade unionist in this group, made it quite clear to Goerdeler, the conservative political leader, that his trade-union network would never take the initiative in an uprising: it had been constructed in order to organize the working class *after* the army had removed the Nazi leadership. Any other strategy would have been, in Leuschner's judgement, simply suicidal.[23] Leber, Haubach, and Mierendorff took a similar view. When they were released from concentration camp in the late 1930s they did not seek contacts with the remnants of the Social Democratic Party underground or the exile leadership, and made no efforts in the direction of independent or

[22] For the basic outline of this remarkable story, Heinz Höhne, *Codeword: Direktor* (London, 1971), can probably be relied upon. A full scholarly enquiry is much needed.

[23] Annedore Leber, *Das Gewissen steht auf* (Berlin/Frankfurt am Main 1956), 97 f.; Moraw, *Parole 'Einheit'*, 53–5, 58.

autonomous resistance activities: they sought out (or were sought out by) political and military allies with immediate access to the levers of power, and tried to establish their own political cause within a resistance coalition.[24]

For their part, the conservative leaders realized that there were no politicians in their own ranks who could appeal to, and thus control, the working class after a successful coup. They need the Social Democrats in this role. This quite novel alliance, however, amounted to a good deal more than mutual need and a shared rejection of Nazi barbarism. The overthrow of Hitler promised these Social Democrats an opportunity to restructure the labour movement and the whole framework of German politics. Thus Leuschner, Kaiser, and Habermann laid the ground in their resistance work for a powerful new unitary trade-union organization, which would fuse the free, the Christian, and the nationalist unions and take control over the future social-insurance system. Leber and his friends were more radical: there would be no return to the party politics of the Weimar Republic, no rebirth of their own party. They would license only a single anti-Nazi political organization, the 'popular movement' (perhaps to be called Socialist Action), which would be place firmly under their own leadership.[25]

One of the motives which drew Leber, Mierendorff, Haubach, Maass, and other like-minded Social Democrats into the military conspiracy was their distinctive sense of the complete inadequacy of their own party in the years before 1933. In a bitter analysis of its history which he wrote while in prison in 1933, Leber pronounced the old Social Democratic Party dead; the defeat was such as to require a completely fresh start.[26] He and his collaborators had always been outsiders in their party, and for much the same reasons in each individual case. They were all relatively young, born in the 1890s; they all served with distinction in World War I and retained a familiarity with military matters and with the military style which was uncharacteristic of Social Democracy. Only Leber came from a proletarian background. All of them completed university degrees, mostly doctorates; they were intellectual idealists, often with a background in the youth movement, who regarded Marxism as at best obsolete. They all entered Social Democratic politics near the top in the

[24] Hoffmann, *Resistance*, 103, 122, 130 f.

[25] Moraw, *Parole 'Einheit'*, 55–62.

[26] The text of this powerful document, 'Die Todesursachen der deutschen Sozialdemokratie', has now been published in full by D. Beck and W. F. Schoeller in *Julius Leber, Schriften, Reden, Briefe* (Munich, 1976).

1920s, without serving a long apprenticeship in the party machine or the unions. Their ambition to lead the party was clear. While their political pragmatism placed them on its right wing, their activism was highly uncongenial to the cautious older reformist functionaries who led the SPD during the Weimar Republic: Haubach and Mierendorff, for example, threw themselves into the grass-roots fight against the Nazi party after 1930, trying to rejuvenate and modernize Social Democracy as a militant mass-movement, and all these men argued in vain for an uncompromising defence of the Republic in 1932–3, if necessary by violent means. Leber was prepared to split the party if this could have given the Republic a strong coalition government—an early signal, perhaps, of the alliances of the resistance struggle. The language of their politics constantly invoked leadership, pleasure in power and responsibility, the readiness to take big risks, the role of the creative personality and of the strong state—motifs which were largely alien to the defensively minded bureaucrats who ran the party after Bebel's death. They were part of a self-conscious political élite of a new generation, who sought new solutions both to the old problems of democracy and the labour movement and to the immediate problem of resistance to Nazism.[27] By 1944 the new trade unions and this new Social Democracy had, if only as leadership groups, a strong position in the political alliance behind Stauffenberg.[28]

In the event, all efforts after 1933 were too late. The defeat was so comprehensive that no strategy of resistance could make it good. The political thinking of the Social Democratic participants in 20 July 1944 points us back, in conclusion, to the reasons for the defeat of the Left and to the reasons why there was no open fight against Nazism.

The leaders of the Communist Party radically underestimated the Nazi movement in the years 1928–33. On grounds which had much to do with the interests of Soviet foreign policy they labelled as fascist all political forces to the right of themselves. While many Social Democrats tended, for propaganda purposes, to dismiss the Nazi movement as a contemptible and unstable political front for reactionary economic interests, the debate within their party about the nature of the Nazi threat was much more serious and realistic. It was the very danger of this threat which justified the party in demanding heavy sacrifice and strong discipline

[27] See the acute biographical sketches of these men by Ger van Roon, *Neuordnung im Widerstand* (Munich, 1967), 41–7, 100–8, 123–31, 181–8, 204–9.

[28] Ibid. 226–34; Hoffmann, *Resistance*, 198–202, 357 f.

from its supporters by tolerating the authoritarian and deflationary policies of the Brüning government.[29] Even more prescient of the terrible power of a Nazi dictatorship were the warnings of left-wing independents like Miles and Trotsky.

The impotence of the Left stemmed less from misperceptions of the situation than from the basic conditions of political action and from the tactics of the struggle for state power. The basic conditions were in large part set by the economic depression, which magnified the disunity of the Left, helped to disarm Social Democracy, and gave a new but highly brittle strength to the Communist Party. The history of the crisis years is too complex to permit a brief summary here, but it is perhaps worthwhile trying to single out in a schematic manner four of the main reasons for the inability of the labour movement to offer effective resistance to the rise of Nazism. First, the Social Democratic Party and the trade unions were thrown comprehensively on to the defensive. The economic crisis robbed the party both of its platform (welfare reforms) and of its political strategy (making alliances to the right). Tolerating the Brüning government at the Reich level, and implementing welfare and expenditure cuts in those states and cities where they held power, were for Social Democrats the demoralizing politics of the lesser evil. The erosion of their electoral support further unnerved the party leadership. The loss of forty-four seats in the Prussian elections of April 1932 was an especially heavy blow, since it deprived the party of its democratic right to power in its last major stronghold. Social Democracy was committed to the ballot box and its parliamentary leaders reacted to electoral defeat by becoming more and more defensive, more cautious about the possibility of extra-parliamentary action to defend the constitution. Rapid political decline was bad ground from which to launch a risky counter-offensive.

Second, partly because the posture of the Social Democratic Party remained so defensive, the Communist Party's attack on it as 'social-fascist' was partially, and disastrously, successful. The Communist Party made sufficiently large electoral gains at the expense of Social Democracy to weaken and embitter the reformist left, but it did not increase its own power enough to constitute the main, or even a serious,

[29] Rudolf Breitscheid developed this argument very powerfully in his address to the 1931 SPD Congress, now reprinted in Wolfgang Luthardt (ed.), *Sozialdemokratische Arbeiterbewegung und Weimarer Republik. Materialien zur gesellschaftlichen Entwicklung 1927–1933* (Frankfurt am Main, 1978), ii. 326–54. This critical edition of texts is a rich source for the history of the reformist labour movement.

obstacle to a Nazi take-over. Its brief tactical alliances with the Nazi party weakened the republican constitution, but did not bring the utterly unrealistic goal of a Soviet Germany one step nearer. Between 80 and 90 per cent of its members were unemployed, and the party led a volatile and ill-organized movement of social protest. Its calls for general strikes were pure gesticulation, designed to discredit the cautious trade union leaders, and it had made no serious preparations for an armed uprising or even for an armed defence of its own organizations. In February 1933 the Communist Party was not strong enough to launch its own attack on the new government: it challenged the reformist left to do so. In this form the bitter conflict between the two working-class parties immobilized both of them. The conflict left no space for the development of new strategies. Mutual distrust was complete.

Third, a real power to act did still lie with the Social Democratic Party, the unions, and their paramilitary defence force, the *Reichsbanner*. The strength of the unions, however, had been badly undermined by mass unemployment, for two-thirds of their members were either out of work or on short time, and the unions now had to face a government strike-breaking organization with large numbers of potential volunteers. A general strike without some kind of simultaneous insurrection did not seem likely to have a decisive political impact: in this respect the constellation of forces was much more threatening, much less controllable, than it had been in 1920 against the Kapp Putsch. A small minority of leaders favoured an offensive strategy, but the unity and security of the reformist labour movement seemed best served by caution. For all these reasons the guiding imperative of Social Democratic politics was a negative one: do nothing which might provoke the reactionary or the Nazi right to acts of violent repression. But this abstinence, strict legality, did not amount to a policy. After the fall of Brüning in May 1932 such provocations could be manufactured. Thus when Papen removed the caretaker government of Prussia in July 1932, the majority of Social Democratic leaders decided that it was too dangerous and too early to risk everything in a civil war. The risks were indeed great. The unions and the *Reichsbanner*, supported probably by sections of the Prussian police, could well have found themselves fighting against regular army units as well as against the large paramilitary formations of the right, while at the same time being unable to control the actions of the Communist Party in any such confrontation. There would at best have been a period of very bloody confusion.

However, the price of inaction was high: the surrender by Braun and Severing of their power over the Prussian police in July 1932 turned six months later into a catastrophe.

Social Democracy claimed to be guarding against and saving itself for an eventuality which never came about, and which Hitler had long since ruled out as a strategy: a *coup d'État*. Hence the fourth and last consideration: at every step in the final phase of the crisis it was the opponents of the labour movement who possessed and retained the political initiative. They controlled the form and terrain of the conflict, made the tactical choices, and determined the sequence of decisions and events in a way which made it very difficult for Social Democracy to respond adequately. Hitler's installation as Chancellor was formally correct, so the moment to strike had still not come—more calls for organization and discipline. One month later the government used the burning of the Reichstag to decapitate and suppress the Communist Party with ruthless speed. The Social Democratic Party remained, briefly, untouched, and some of its members thought that the Communists only had themselves to blame. Then in March and April Nazi storm-troopers destroyed the unions and Social Democratic organizations at the grass roots, vandalizing offices and killing, beating, and carrying off officials as the police stood by. This too was not a coup or a formal proscription. It proved impossible to fashion adequate or appropriate counter-measures to so unpredictable and formless a wave of terror; the victims were reduced to petitioning the government to restore the rule of law. While many party and union leaders hoped that the outrages would peter out, Hitler and his advisers saw them as a test of strength. When some leading trade-unionists showed themselves willing to try to find a basis for compromise with the government, and at the same time the Social Democratic Party began to split over whether to go underground or not, the regime shut them both down, without difficulty. *Reichsbanner* members tried to hide their weapons.[30]

One feature which stands out in the politics of Social Democracy and the trade unions in these years is the distance between leaders and members. This distance took many forms. Both the party and the unions were highly bureaucratized institutions. Members had little direct influence over the composition or policies of leadership groups; the personal position of the leaders went largely unchallenged through almost three

[30] I have treated the developments in the first months of 1933 in greater detail in 'National Socialism and the Working Class, 1925–May 1933', in *New German Critique*, 11 (1977). See also Matthias, 'Untergang', esp. 203.

years of crisis and precipitous decline. This concentration of power in the hands of the top functionaries gave rise on their part to a distinctive organizational ethic, to an oppressive and impractical sense among the leaders of their own exclusive responsibility for the welfare of the organizations and the members: above all the unity of the organizations had to be preserved intact.[31] As Social Democracy went from one setback to the next defeat this sense of responsibility turned into a paralysing burden. There were no well-developed mechanisms for sharing the burden with the members and supporters who would have died in an insurrectionary general strike. Some, perhaps many, of these people would have preferred to make a fight of it in 1932 or 1933, whatever the outcome. Most of their leaders could not reconcile such a decision with their sense of organizational responsibility. At least one, Otto Wels, wondered whether the rank and file might act decisively on its own, and felt that the members had proved inadequate to the challenge![32] The leaders lacked familiarity with and confidence in the determination, political judgement, and inventiveness of their own supporters, and they attached too little importance to morale. 'I must tell you', one militant wrote in 1934 from his own place to exile to a member of the exiled party committee in Prague, 'that the terrible lethargy of the party executive, its complete lack of initiative, made many old comrades feel lonely. Many went their own ways, embittered.'[33] The character of the defeat in 1932–3 weighed heavily upon those who did not go their own ways, but took up the illegal underground struggle.

[31] Theo Pirker's statement in K. D. Erdmann and H. Schulze (eds.), *Weimar. Selbstpreisgabe einer Demokratie* (Düsseldorf, 1980), esp. 343, is very eloquent on this point.

[32] Helga Grebing, 'Auseinandersetzung mit dem Nationalsozialismus', in Luthard (ed.), *Sozialdemokratische Arbeiterbewegung*, ii. 272; Matthias, 'Untergang', 206.

[33] Otto Buchwitz to Paul Hertz, quoted in Moraw, *Parole 'Einheit'*, 63.

7

ANTI-JEWISH POLITICS AND THE IMPLEMENTATION OF THE HOLOCAUST

HANS MOMMSEN

1. Introduction: The Holocaust and the erosion of morality

THE task of the historian is to provide rational explanations of past developments, to describe their complexity, to draw comparisons with analogous processes, and to analyse their structures. With the Holocaust he seems to be confronted with a unique and incomparable chain of events which transcends ordinary historical consciousness and has a dimension which is fundamentally anthropological. The deliberate murder of more than five million human beings in an advanced Western industrial society in the middle of the twentieth century appears to be an unmistakable warning sign of how thin is the patina of Western civilization covering uncontrolled instincts, egotism, and barbarism, depicted in Thomas Hobbes's account of the state of nature. Far worse, while Western civilization has developed the means for unimaginable mass-destruction, the training provided by modern technology and techniques of rationalization has produced a purely technocratic and bureaucratic mentality, exemplified by the group of perpetrators of the Holocaust, whether they committed murder directly themselves or prepared deportation and liquidation at the desks of the Reich Main Security Office (*Reichssicherheitshauptamt*), at the offices of the diplomatic service, or as plenipotentiaries of the Third Reich within the occupied or satellite countries. To this extent the history of the Holocaust seems to be the *mene tekel* of European progress and civilization since the emergence of the modern state; in this respect Helmuth James von Moltke's idea, cherished also by Adam von Trott zu Solz, that it was necessary to correct the course that Western history had taken since the Reformation was not just utopian speculation.[1]

But the historian, confronted with the task of explaining why the Holocaust could be carried out, is also overburdened by the specific

[1] For Moltke's political philosophy see Ger van Roon, *Neuordnung im Widerstand. Der Kreisauer Kreis innerhalb der deutschen Widerstandsbewegung* (Munich, 1967), 507 ff.; the biographical background is depicted by M. Balfour and I. Frisby, *Helmuth von Moltke. A Leader Against Hitler* (London, 1972).

nature of these events which lack even the traits of a tragedy, which, Jan Huizinga complained, has been lost in the course of contemporary European history. Hannah Arendt's famous formula, the 'banality of evil', received with such controversy, refers not only to the mental and psychological conditions under which the Holocaust was implemented, but also to the process itself, which in many respects lies beyond those categories the historian, especially if he is trained in the idealistic inheritance of historicism, employs to impart structure to amorphous historical facts.[2] Certainly, the chain of events leading to the liquidation of millions of European Jews has been revealed and the circumstances, at least in comparison with other epochs, are well researched. We cannot expect substantially to extend the documentary evidence, except for the procedures in some of the occupied countries and the international repercussions of the Holocaust. But human nature tends to repress the unimaginably cruel and horrifying conditions which accompanied the factory-like mass murder committed in the extermination camps and the massacres committed by the *Einsatzgruppen* or task-units in the East. The language of the historian is inadequate to describe this climax of inhumanity without resorting to over-rationalization.

Under extreme conditions of human depravity there is a weakening of those mechanisms of social conduct which usually guarantee the existence of society. In the eyes of the liquidators, but also of those who accidentally witnessed the selections at the Auschwitz ramp or similar events, the Jews had already lost the character of human beings. They had been deprived of the quality of human individuals and ranked far below the meanest criminal, who at least preserved his personal identity until his execution. The victims who were driven into the gas-chambers and gas-vans did not even have that status. The process of dehumanization included the perpetrators as well. It was not only that they acted immorally; in fact, not even the quasi-moral mechanisms of human reaction were operating any longer. This is the most difficult thing to explain. Many of the perpetrators were not radical anti-Semites, and anti-Semitic racial indoctrination was only one factor among many, if one is to explain why, for instance, Otto Ohlendorff took the job of leading one of the *Einsatzgruppen* and of fulfilling his murderous obligations with bureaucratic perfectionism.[3] It is difficult to assess the

[2] See H. Arendt, *Eichmann in Jerusalem. Ein Bericht von der Banalität des Bösen* (Munich, 1954), 188 ff.

[3] See H. Krausnick and H.-H. Wilhelm, *Die Truppe des Weltanschauungskrieges. Die Einsatzgruppen der Sicherheitspolizei und des SD 1938–1942* (Stuttgart, 1981).

fact that on a certain level the liquidation of human beings appeared as a rather ordinary job which was performed without any sense of ideological or moral involvement.

This does not mean, however, that the perpetrators, from the top down to the bottom, did not need some sort of moral or pseudo-moral legitimation of the genocide they practised. It seems to be symptomatic that the storm-troopers already involved in the pogrom of November 1938, asked for the reasons behind their criminal conduct, would not argue on anti-Semitic grounds, but justified their deeds by pointing to their loyalty to their leaders, obedience to orders, and similar arguments of a secondary character. As Raoul Hilberg has proved, the infamous reports by the *Einsatzgruppen* also contained attempts to justify the liquidation of Jewish people not directly related to anti-Semitic racial ideology.[4] Liquidations for which a comparable pretext could not be found did not take place. The anti-Semitic slogans notoriously spread by the National Socialist press (not only by organs like *Der Stürmer*), demanding the elimination of the 'Jewish race', were apparently received by the broader public as rather meaningless propaganda metaphors, unrelated to everyday experience. Unquestionably, the *Reichssicherheitshauptamt* from the start had developed a disguised language to protect the secrecy of the liquidation process, and (what should not be overlooked) not to disturb the technocratic efficiency of its implementation. Even the victims preferred to cling to this indirect terminology, and in doing so they reacted in a similar way to the perpetrators. Only by partially repressing the bitter and intolerable reality could the destruction process be performed.

Hence, we discover an overall inclination to disconnect anti-Semitic racial propaganda from what really went on in the extermination camps. Furthermore, a split in moral self-perception appears to be typical not only of members of the SS, who were eager to develop a pseudo-morality as regards genocide, but also of the Nazi system in general. It is well known that perpetrators of the Holocaust like Rudolf Höss, the commander of the concentration camp of Auschwitz-Birkenau, combined their ordinary function of organizing death with a private sphere characterized by petty-bourgeois morality and the deliberately upheld fictional world of secondary values such as order and cleanliness, the fulfilment of one's duty, and phrases like 'honour', 'service to the

[4] See Raoul Hilberg, *Die Vernichtung der europäischen Juden. Die Gesamtgeschichte des Holocaust* (Berlin, 1982), 166, 223 ff.

nation', and 'loyalty'.[5] Heinrich Himmler's speeches of October 1943 and 1944 in which the *Reichsführer SS* disclosed the terrible secret in order to lay the responsibility for it on broader shoulders reveal the double morality symptomatic of Nazi rule.[6] Although the liquidations were frequently accompanied by criminal and even sadistic acts, Himmler claimed that his men, by performing the 'heaviest task' the SS ever had to bear, without having lost their moral countenance, had remained *anständig*, a German word so imbued with connotations of an eroded bourgeois morality that it is almost untranslatable. Obviously, pseudo-moral justifications as represented in Himmler's disclosures of genocide policy cannot be classified as primary motivations for the fundamental step towards technocratic mass-extermination, although they were intermingled with anti-Semitic racial prejudices.

No interpretation of the Holocaust can omit the question why no significant opposition to the deportation and liquidation processes emerged. In fact, recent research has established that the programme, despite its formal secrecy, involved groups far broader than the small units drawn from the SS Death's Head organizations who were in charge of the concentration camps and the killing operations. Undeniably, the armed forces were actively involved, not only because the leading generals had participated in drafting the complex of criminal orders, but also because ordinary troops in many parts of the German sphere of influence had been indispensable either in preparing the deportations or in creating the preconditions for immediate mass killings.[7] Moreover, the German railways, the Reichsbank, the civil administration in the occupied territories, and the police forces (sometimes non-German in nationality) contributed to the implementation of the 'Final Solution' in many ways. The burning question is whether they did not know what they were really doing or whether they proved to be psychologically

[5] See the introduction by Martin Broszat to the memoirs of Rudolf Höss; *Kommandant in Auschwitz. Autobiographische Aufzeichnungen von Rudolf Höß* (Stuttgart, 1958).

[6] In a few secret speeches Himmler disclosed the systematic genocide measures to Nazi leaders and other high-ranking officials, thereby partly breaking the secrecy of the liquidation policies. The speeches were delivered on 6 Oct. and 16 Dec. 1943, 24 May and 21 June 1944. They are edited by F. Smith and A. Peterson, *Heinrich Himmlers Geheimreden 1933 bis 1945 und andere Ansprachen* (Frankfurt, 1974) 201 ff.

[7] See H. Krausnick, 'Kommissarbefehl und Gerichtsbarkeitserlaß Barbarossa in neuer Sicht', *Vierteljahreshefte für Zeitgechichte*, 25 (1977) 716 ff.; Christian Streit, *Keine Kameraden. Die Wehrmacht und die sowjetischen Kriegsgefangenen 1941–1945* (Stuttgart, 1978), 51 ff.

unable rationally to comprehend the reality behind the regime's disguised language.

2. *Cumulative radicalization of anti-Jewish persecution until World War II*

To many observers anti-Jewish policy and genocide seemed a necessary consequence of Nazi ideology and Hitler's anti-Semitic tirades in *Mein Kampf.* Doubtless, racial anti-Semitism appears to be the most stable and continuous element within the generally incoherent and never clearly elaborated ideological programme of the Nazi party. It is well known that Hitler in particular clung unremittingly to his anti-Semitic clichés up to the very end in the *Führerbunker*. Hence, the question arises why the anti-Semitic part of Nazi ideology had or attained top priority among other goals of the Nazi movement. There is an ideological explanation for this, in so far as anti-Semitism provided a link between the different and incoherent elements of Nazi ideology, which primarily reflected dominant social resentments within society and were, at least from Hitler's viewpoint, used above all as a tool for political mobilization. However, the crucial function of anti-Semitism was to bridge the objective conflict between the anti-capitalist and anti-socialist tendencies of the movement. This, however, does not sufficiently explain why the Nazi regime gave the deportation and later on the extermination programme a top priority in comparison with other programmes which were postponed, especially since the anti-Jewish policy continually came into conflict with the immediate military, economic, and diplomatic goals of the regime.

During the first years of the Nazi dictatorship foreign and domestic policy compelled the regime to moderate the anti-Semitic outbursts of the otherwise politically underemployed SA units. Until 1938 economic considerations put forward by Schacht and others had as their consequence that the long-demanded 'Aryanization' of Jewish economic property was not backed by legislation, but rather was promoted on an unofficial level. A great many of the semi-spontaneous anti-Jewish actions early in 1933 carried out by local SA units apparently threatened the Nazi alignment with the conservative élites and proved to be disadvantageous for foreign policy. Most significantly for him, Goebbels tried to overcome the government's embarrassment by suddenly starting a 'flight forward' through his active involvement in Streicher's boycott in April 1933. The boycott had to be prematurely cancelled; the population showed its opposition by demonstratively buying in Jewish shops and department-stores during the following days. Goebbels had expected that the masses would support the party action and that the latter would lead to pogrom hysterias; the opposite was true. Goebbels recognized that the Nazis would have to go the gradual path of progressing pseudo-legal measures against the Jews.

In fact, anti-Semitic feelings were virulent only in certain sectors of the German society; on the whole radical anti-Semitism was quite limited, particularly in comparison with East European countries like Poland. Membership of radical anti-Semitic organizations during the Weimar Republic did not exceed 250,000; only some of these people found their way to the NSDAP before 1933. It was significant that specific anti-Jewish orders like the prohibition on buying goods in Jewish shops had to be repeated several times. Even many of the technicians who later on ran the extermination machinery were originally not deeply rooted anti-Semites, as is shown by the example of Höss, the commander of Auschwitz, or Eichmann, the organizer of the 'Final Solution'. Both eagerly picked up the prevailing anti-Semitic slogans and prejudices to serve as an additional motivation of their rather bureaucratic performance. It took several years before Nazi propaganda succeeded in spreading an extremely unfavourable stereotype of the 'Jew' modelled on the non-assimilated 'Eastern Jews', those Jewish groups who migrated to Germany after 1918, especially after the Option Treaty with the Polish Republic in 1921 provided the opportunity of acquiring the German citizenship. The ardent and zealous anti-Semitism of the Nazi leaders, particularly Himmler, Heydrich, Bormann, Goebbels, and Streicher, was not necessarily adopted by the rank and file of the party. The working class was, in accordance with its specific political traditions, almost completely free from any anti-Semitic 'resentments'. Socio-economic interests were primarily involved for small businessmen, artisans, and shopkeepers, who felt the competition of Jewish department and discount stores and who supported 'most enthusiastically' the exclusion of Jews from the German economy. Competition in economic areas numerically dominated by Jews was frequently regarded as an 'objective' cause for anti-Semitism. A closer socio-economic analysis, however, shows very clearly that the previous overproportionate representation of Jews in certain economic sectors, particularly in banking and trade, was, independent of anti-Semitic propaganda, already declining as a consequence of the general structural changes of the capitalist system. From the economic viewpoint there had been no real 'Jewish Question' since the turn of the century.

These facts make it plain that anti-Semitism could not serve as an overall integrating ideology to attain the unity of the political will of the German people that so many longed for. On the other hand, there can be no doubt that particularly among conservative élites and the official churches, latent anti-Semitic 'resentments' led to a deplorable moral indifference; university professors, in particular, did not launch any open and energetic protest against the dismissal of their Jewish colleagues. But there was also a considerable amount of readiness to help persecuted Jews above all in the big cities, and this must be weighed against the background of sharpening political pressure on anybody who did not conceal his sympathies for Jewish citizens. Nevertheless, the question arises why a majority among the population tolerated anti-Jewish actions, either terrorist measures by SA units or central legislation. We know very well that the Nazi system was fairly vulnerable to open public protests. The example of euthanasia proves that public action could compel the regime to change its policy; the intervention by the churches forced Hitler to stop the infamous 'Action T4' (the killing of mental patients, so called because of the office from which it emanated, Tiergartenstraße 4), which was later resumed on a limited scale in the concentration camps. The problem why no comparable protests occurred against the persecution of Jews raises a variety of underlying questions, which refer primarily to the political strategy pursued by the Nazis with regard to the Jewish minority.

In hindsight, the Nazi policy against German, then European, Jewry seems to be a consistent sequence of deliberate measures which necessarily culminated in the Holocaust. The many interpreters of Nazi anti-Jewish policy who adhere to this explanation usually argue that the extermination programme was expressly laid down in Hitler's *Mein Kampf*. That Hitler, like his anti-Semitic forerunners, thought or speculated about extinguishing the Jewish 'race' by violent means is undeniable. But this does not mean that Nazi policy necessarily had to lead to the killing of roughly five million Jewish people. Even the most fanatical anti-Semites within the Nazi élite or the inner circle around the dictator did not conceive the 'Jewish problem' from the aspect of a possible extermination policy before the end of 1939, and even then there was no clear conception of what could be done against the Jews. One could argue—and many historians have done so—that the Nazis successfully concealed their true intentions, aware that any extermination programme was not to be realized within a short period, but would require at least several years. Hitler's warning of 30 January 1939, 'If international finance Jewry outside Europe succeeds in propelling the German people again into a World War, the result will not be Bolshevization of the earth and thus the victory of Jewry, but the extermination of the Jewish race in Europe',[8] was not taken at face value, either by foreign observers or by the Nazi power-technicians primarily involved in the political solution of their self-generated Jewish question.

Heydrich and Adolf Eichmann, who, first in Austria and then in the Reich, acted as chief organizers of Jewish emigration, thought in more realistic categories and did everything they could to foster Jewish emigration. Most significantly, Heydrich favoured the Zionist groups among German Jewish organizations, and originally Eichmann was in touch with Zionist committees in Palestine. It belongs to the inherent and most significant internal contradictions of the Nazi system that Eichmann became more and more reluctant to face the fact that the creation of a Jewish community in the Near East would produce an embittered enemy of Germany in the long run. Therefore this scheme was dropped, but emigration in general was expanded in several ways.

The main question relates to the circumstances under which an obviously unrealistic and fictitious programme was turned into an operational 'political goal' that received priority even over the immediate military objectives of the regime. Martin Broszat has characterized this process as 'a negative selection of the elements of the National Socialist *Weltanschauung*';[9] it can be analysed with particular reference to the anti-Jewish policies of the Third Reich. The example of the origin of the Nuremberg Laws reveals the continuous radicalization of anti-Jewish measures as a partial outcome of ambivalent decision-making in the Third Reich.

The Nuremberg Laws consisted in the Reich Flag Act, the Reich Citizens Act, and the so-called Law for the Protection of German Blood

[8] Hitler's speech is printed in M. Domarus, *Hitler. Reden und Proklamationen*, ii. 1 (Munich, 1965), 1057 ff. He quoted this passage in his speeches of 30 Jan. 1941, 30 Jan. 1942, and 8 Nov. 1942, but there are similar utterances as early as in Apr. 1933; compare my systematic evaluation, Hans Mommsen, 'Die Realisierung des des Utopischen: die "Endlösung der Judenfrage" im "Dritten Reich" ', *Geschichte und Gesellschaft*, 9 (1983), 395 ff.

[9] See the excellent analysis of Hitler's role within the Nazi regime by Martin Broszat, 'Soziale Motivation und Führerbindung des Nationalsozialismus', *Vierteljahrshefte für Zeitgeschichte*, 18 (1970), 400 ff.

and German Honour. They were ceremonially promulgated at a special session of the German Reichstag held at Nuremberg in connection with the Reich Party Convention of Freedom on 15 September 1935. Contemporary observers as well as later historians regarded the Nuremberg Laws—particularly the Blood Protection Act, which contained a series of discriminatory measures against Jews—as a decisive interruption in German domestic development and as an intensification of the anti-Jewish policies of the German government. They were regarded by the Western world as a triumph of the radical wing within the NSDAP. The domestic reaction was quite different. This was partly due to the impact of the official statements claiming that the new legislation aimed at a clear-cut separation of the Jewish from the majority population and provided the opportunity for German Jews to become a national minority. Interpretations of this kind were guided by the endeavour to avoid further protests outside Germany which could threaten the Olympic Games scheduled for Berlin in 1936. But apart from that propaganda manœuvre, the Nuremberg Laws were evaluated even by Jewish groups in Germany (in spite of their discriminatory character) as a principal step towards a legislative solution of the hitherto undecided question of the future social status of Jews in Germany. The Nuremberg laws did not contain any measures concerning Jewish Germans' economic position except for the expectation that they would further restrict Jewish access to the civil service.

The real story of the Nuremberg Laws and what happened behind the propaganda façade built up by Frick, and partly by Goebbels, was a different one. It was unusual for a government empowered by the Enabling Act to rule by governmental ordinances or decrees to choose the platform of the Reichstag to proclaim laws which many party functionaries and Nazi officials held to be overdue. The promulgation of the laws through the Nazi Reichstag, however, was nothing else than the successful attempt to replace the Reichstag's original agenda. On the advice of Neurath, Hitler had called the Reichstag for a separate session to Nuremberg, a decision he took three days before the party convention was to start. Hitler intended to deliver a foreign policy speech in the presence of the deputies and the Diplomatic Corps specially invited to attend the Reichstag session, which should provide the propagandistic scenery for him to address the great powers and to make a proposal for achieving an agreement on the Abyssinian question. On the morning of 13 September Hitler dropped this plan on account of a change in Mussolini's policy. The question arose what to do with the Reichstag.

The Reich Flag Act proposed by Frick, which finally gave the swastika a monopoly as the state symbol, appeared to him to be too meagre in view of the extraordinary organizational and propagandistic efforts connected with a special Reichstag session in Nuremberg.

The day before the party convention opened, the secretaries of state in the Reich Ministry of the Interior informed their subordinates responsible for racial policy and naturalization legislation that Hitler wished to get an immediate draft for an act covering the pending questions concerning the legal status of Jews. In consequence, the Nuremberg laws were drafted without taking into account earlier legislative preparations in the Ministry of Justice, although it was certainly involved in the issue of so-called 'racially mixed marriages'. No wonder that the Minister of Justice, Gürtner, had no knowledge of the new legislation before the session of the Reichstag, and the same was true for most of his colleagues.

The responsible administrative expert, Bernhard Lösener, described after 1945[10] the almost unbelievably hectic and inadequate conditions under which officials of the Interior Ministry prepared different drafts of the requested *Judengesetz*; apart from the extreme time-pressure, the civil servants were confronted with the fact that Hitler steadily changed his mind owing to the influence of party leaders surrounding him, particularly the leader of the Nazi Doctors' Association, Gerhard Wagner, who represented the ultra-radical views of the party mandarins. The conflict primarily turned on the question whether already existing mixed marriages should be dissolved and whether one-eighth-Jews should be exempted. The same dispute arose over the Reich Citizens Act which announced, in addition to existing nationality laws, a separate Reich citizenship from which Jews were to be excluded, becoming *Schutzverwandte* of the German Reich and thus attaining a clearly underprivileged legal status. Politically important during the Nuremberg negotiations was only to what extent the 'Aryan paragraphs' of the Nazi party statutes should become part of official Reich legislation.

[10] Bernhard Lösener, 'Als Rassereferent im Reichsministerium des Innern', *Vierteljahrshefte für Zeitgeschichte*, 9 (1961), 264 ff. Although Lösener's report in general is reliable, it is marked by apologetic elements stressing too much the oppositional role of the bureaucracy. Compare my remarks in 'Realisierung des Utopischen', 387 f. and Lothar Gruchmann, '"Blutschutzgesetz" und Justiz. Zur Entstehung und Auswirkung des Nürnberger Gesetzes vom 15. September 1935', *Vierteljahrshefte für Zeitgeschichte*, 31 (1983), 418 ff.

The background of the disputes between the Ministry of Interior, particularly the civil servants (because Frick showed no intention of backing his subordinates, but, at the same time, hesitated to dismiss their technical knowledge), and the representatives of the radical wing of the Nazi party consists in the phenomenon that the NSDAP complained bitterly that they were being neglected and overruled. The speeches at the party convention show very clearly the sharp tension between the party leadership and the ministries' bureaucracy. The party functionaries were convinced that the 'Jewish question' should be handled by the party alone and not by those contemptible bureaucrats who, as a close adviser of Bormann in the Deputy Führer's staff put it, wanted to establish an absolutistic state of their own.[11] But the party obviously did not succeed in getting permission for free action. The strong desire within the party organization to carry the anti-Semitic measures into the economic field were unanimously rejected by the Reich cabinet, including Heydrich, when it met on 20 August at the request of Schacht. Only the Reich Minister of Propaganda sided with the representatives of the Deputy Führer, who asserted that about 70 to 80 per cent of the population wanted more radical steps agains the Jews. In a secret meeting of *Gauleiter* at the end of the Nuremberg party rally, Hitler himself repeated the ban against interference in Jewish enterprises and against illegal methods. The party leadership was already suspicious what the special Reichstag session might do, and, most significantly, Goebbels informed the secret Reich Press Conference, the compulsory meeting of the German journalists who formed the Reich Press Chamber of the Ministry of Propaganda, that there would be no statements on domestic policy, and no treatment of the Jewish question. Under these circumstances it is not surprising that the party regarded the Nuremberg Laws primarily as a defeat, and the reaction in party circles was clearly negative. This explains also why Goebbels gave orders to interrupt the radio transmission of the Reichstag session immediately before the three Nuremberg Laws were proclaimed and that he restricted the reports in the newspapers to reprints of the official commentaries.

Goebbels and the radical wing of the party leadership hoped to reach their goals in the secret leadership meeting in Munich scheduled for 24 September. In contrast to their expectation, Hitler, after delivering a four-hour speech, eventually accepted the more moderate proposals of

[11] Compare Hans Mommsen, *Beamtentum im Dritten Reich* (Stuttgart, 1966), 114 f.

the ministries' bureaucracy and decided that half-Jews who did not belong to the Mosaic religion were to be exempted from the new regulations.

The origins of the Nuremberg Laws show that the controversies between the radical groups in the party leadership and the bureaucrats meant that the more extremist goals of the party were not implemented until the war years. Indirectly, the bureaucrats' limited success and their moderating influence contributed to preparing the foundations and preconditions for the later extermination policy. The inclusion of persons of mixed racial descent up to one-eighth-Jews would have involved broad strata of the assimilated Jewish population in the measures of suppression and necessarily would have touched a very large number of Gentiles, even supporters of the regime. It was only on account of limiting the definition of the group to be provisionally exempted from direct persecution that Heydrich's policy of isolating the Jewish minority socially and morally from the majority population proceeded without major protests from the public, which was not really aware of what was going on or tried to repress any information about it, because that part of the Jewish population who had been in close contact with their German neighbours were either not included in the growing discrimination or were step by step isolated from them. Only after cumulative discriminatory legislation, in whose formulation various ministries competed in order not to lose their influence, had pressed Germany's Jews into the role of social pariahs, completely deprived of any regular social communication with the majority population, could deportation and extermination be put in effect without shaking the social structure of the regime. There existed a dialectical relation between the anti-radical, moderating initiatives of the bureaucracy, particularly in reference to the 'Aryan paragraph' regulations, and the spontaneous wishes of the party functionaries who wanted to get the Jewish question as their specific field of action without possessing any clear-cut conception of what the outcome of anti-Jewish discrimination would be. This dialectic was speeding up the process of anti-Jewish persecution.

It is very clear that intra-party rivalries as well those between the party and state agencies contributed decisively to what can be described as a process of cumulative radicalization. Interestingly, the fiction of achieving a final legal solution which would provide for the Jews in Germany the status of a national minority even under the most discriminatory conditions prevented the rather conservative officials in the Ministries of Interior and Justice from refusing further collaboration. After 1941, the bureaucracy tended to get rid of any formal responsibility for anti-Jewish legislation. Hence, they indirectly supported Himmler's strategy to extend the responsibility of the Gestapo and to exclude the judiciary and the Ministry of Interior from handling Jewish affairs, which appeared merely as a police problem. The pogrom of 9 November 1938, and the legislative measures proclaimed thereafter, again show the impact of competing interest groups on the political decision-making process. It is well known that Goebbels backed the whole action in order to improve his weakened position and, at the same time, to improve rather poor party finances by gaining control over Jewish property. Goebbels later on claimed to have acted at the preceding meeting with the approval of the dictator of the Old Fighters in the Bürgerbräu cellar, but there are indications that Hitler did not realize the organized character of the pogrom and its extension. There is no doubt that the action was very energetically condemned by Himmler and Heydrich for being detrimental to their emigration policy, not because of the illegal procedures and their extreme immorality. Conversely, they immediately exploited the event for their own ends, especially by taking into custody about 20,000 wealthy Jews who were compelled to hand over their property in order to provide the lacking funds for emigration of propertyless Jewish families.

In fact, the competing interests of the party, represented by Goebbels, the SS, represented by Himmler and Heydrich, and the Four Years Plan under Göring could only be reconciled in the perspective of further radicalization. As a consequence, the 'Aryanization' of Jewish economic property now became legalized, after it had been illegally practised on a large scale in the annexed Austrian territory, where, in accordance with the model set by Streicher in Franconia, the party functionaries enriched themselves by taking over large shares of Jewish property, particularly that of the great many Jewish shopkeepers in Vienna.[12]

The systematic expropriation of German Jewry which was then the order of the day had consequences which were possibly envisaged by Heydrich, but were certainly not perceived from the viewpoint of future extermination. After Jewish emigration became more and more difficult, partly through lack of foreign currency, partly through the stiffening reaction of potential host countries, partly because many active and young individuals had already left, the social situation of the Jews, in spite of the heroic efforts of the Jewish associations, which were forced

[12] See the survey given by Uwe Dietrich Adam, *Judenpolitik im Dritten Reich* (Düsseldorf 172), 204 ff.

to merge in the *Reichsvereinigung* used by Eichmann as a instrument of 'control' and organizational tool, created an almost insoluble problem because more and more Jews were near to starvation. One has to bear that in mind in order to understand the ensuing persecution, culminating in the systematic deportation of German and European Jewry.

3. *The role of Hitler*

All explanations of the origins of the Holocaust focus on the crucial importance of Hitler's personal contribution; sometimes one gets the impression that the genocide above all is attributed to Hitler in an almost monocausal way. That his fanatical hatred of Jews was of decisive importance for Hitler's general political conduct is unquestionable. But it seems that Hitler perceived the 'Jewish question' primarily in propagandistic terms and in a specific visionary context; he did not show much interest, and certainly no active involvement, in the individual steps of the Nazi anti-Jewish policy. In his attitude to this as to other political matters (except for foreign policy and, after 1941, in military questions) Hitler would intervene occasionally, and then in a quite unsystematic way, but in general he would avoid committing himself directly to anti-Jewish actions, especially since he was aware that these were received rather negatively by the German public.[13] Furthermore, in the vast majority of those cases in which Hitler was confronted with the need to establish priorities between different aspects of policy—as in the question of half- and quarter-Jews—he tended to pursue a rather cautious, sometimes even defensive line and to support the more moderate views of the ministerial bureaucracy versus the radical demands articulated by party officials. Neither the boycott action of April 1933, nor the Nuremberg Laws of September 1935, nor the pogrom of November 1938 was instigated by an initiative from the dictator. The pattern of the cumulative radicalization of Jewish persecution until 1940 is one of unco-ordinated actions, which were started by radical anti-Semitic groups within the NSDAP and resulted in a cumulative political pressure on the ministries, which demanded to retain (as the Minister of Interior, Wilhelm Frick, put it) 'the leadership in the Jewish question' for themselves.[14] This resulted in the typical phenomenon that in order to

[13] See Hans Mommsen, 'Realisierung des Utopischen', 387 f.; Adam, *Judenpolitik*, 163 ff.

[14] Chief conference in the Reich Ministry of Economics, 20 Aug. 1935 (Handakten Lösener, F 71/2, IfZ.)

stabilize its continuously weakening position within the political system, the bureaucracy felt compelled to make material concessions, apart from the fact that the establishment of special offices for the treatment of Jewish affairs in the ministries produced a self-propelling process of increasingly discriminatory legislation without much interference from the party.

Hitler never sympathized with the course of excluding the Jews from social life by legislative means pursued by Goebbels and the SS, who thought an allegedly 'orderly' procedure would produce less public disturbance and fit in the prevailing public confidence in the principle of formal legality. Hitler would give his approval to individual, so-called 'wild' actions and would defend even brutal crimes performed in conjunction therewith, but he was not attached to an orderly legislative process of eliminating the Jews from German social and political life, the aim of the bureaucrats who sacrificed the principle of justice to save formal legality. The political process which after 1938 resulted in a centralization of responsibility, formally in the hands of Göring and actually in those of Heydrich and his assistant Adolf Eichmann, was not primarily fostered by Hitler himself. It was symptomatic that the dictator, although strongly supporting activities against the Jews, was not greatly interested in the results of Heydrich's and Eichmann's efforts to promote Jewish emigration. In his meeting with Admiral Horthy in the summer of 1943, for instance, Hitler, who was otherwise known for his interest in detailed statistical data, still used the demographic figures of the early thirties in order to describe the 'Jewish question' in Germany.[15]

As a consequence of this, the escalation of anti-Jewish measures, leading towards a complete social segregation of Jews and (because of their expropriation) making them social pariahs, came into being not as a result of a single plan but as a consequence of a combination of unco-ordinated strategies. While Hitler supported this process ideologically, he never designed it and it was symptomatic of his reactions that he still believed in the necessity of raising special Jewish taxes when the Jews themselves were already the victims of the systematic expropriation process especially fostered by Göring, who hoped thereby to improve the financial resources for German rearmament.

[15] Meeting of Marshal Antonescu on 13 Apr. 1943, with Admiral Horthy on the 16th and 17th of the same month (see Andreas Hillgruber, *Staatsmänner und Diplomaten bei Hitler* (Frankfurt, 1970), 232 f., 245, 265 f.

Many authors still claim that Hitler had already made up his mind during the early twenties to seize the first political opportunity of achieving the liquidation of the Jewish population. But the thesis that Hitler acted according to a long-established plan implies that he adopted, in general, a more moderate position for purely tactical reasons so as to mediate between conflicting targets of the party radicals and the state administration. Such an attitude, however, does not fit into the typical pattern of Hitler's decision-making. The dictator tended to postpone decisions until general conditions made them overdue and possible alternatives no longer existed. Most significantly Hitler did not devise the aforementioned disguised language and would not admit any knowledge of, or involvement in, the actual genocide measures. Simultaneously, he preferred all the time an extremely fanatical language against the Jews, who appeared as subhuman and dangerous vermin.

Gerald Fleming, who has analysed Hitler's responsibility for the Final Solution in detail, has reached the conclusion that the dictator in fact avoided any direct identification with it; only to this extent does David Irving's contention that even in the inner circle this matter was never mentioned in any direct way appear to be correct.[16] Hitler's famous, often-repeated utterance that in the event of war German Jewry and even world Jewry would be extinguished was first made in April 1933 and was understood even by the leading representatives of the regime as rhetoric, designed to cast the Jews in the role of hostages so as to compel the Western powers to give in to Germany's foreign policy aims. It is difficult to prove anything but indirect utterances which referred concretely to the ongoing liquidation. Even the talks with Marshal Antonescu and Admiral Horthy must be considered as examples of Hitler's typical propaganda-metaphors.[17] Hitler's later utterances, his secret speech to generals and officers of the army in May 1944, and his Political Testament, also confirm his habit of referring only in vague ideological and metaphoric terms to the Holocaust.[18]

[16] Compare Gerald Fleming, *Hitler und die Endlösung* (Wiesbaden, 1982), 155, 163 ff., 168 ff.; David Irving, *Hitler's War* (London, 1977), 391 f. Irving's far-reaching conclusion that Hitler did not receive information about the liquidation programme before Oct. 1943 and himself postponed the solution of the Jewish question is not acceptable; see Martin Broszat, 'Hitler und die Genesis der "Endlösung" ', *Vierteljahrshefte für Zeitgeschichte* 25 (1977), 739 ff.

[17] See above, n. 15 and Mommsen, 'Realisierung des Utopischen', 392 f.

[18] Secret speech on 26 May 1944; in Pers. Stab RFSS, IfZ, MA 316, Bl. 4994 ff.; H. R. Trevor-Roper, (ed.), *Le Testament politique de Hitler* (Paris, 1959).

Fleming's conclusion that this is to be explained by Hitler's intention to mask his personal responsibility seems to be plausible, if we could trace any concrete statement on this question within the inner circle of the main perpetrators.[19] Most significantly, Himmler refers to Hitler's speech of August 1939 in his aforementioned semi-public utterances regarding the mass liquidations already carried out; in conjunction with this Himmler mentions a military command of Hitler's which is obviously identical with the *Kommissarbefehl* issued by Hitler in June 1941 as part of the preparations for the war of racial annihilation against the Soviet Union. To this extent, the Himmler speeches do not prove the existence of any distinct order in relation to the extinction of the European Jewry. The only two concrete orders consist in the complex of criminal orders issued in June 1941 for the systematic liquidation of Bolshevik *politruk*s and potential opponents among the civilians, then used for the murder of the Jewish population, and in the later order to intensify the anti-partisan war in 1943.

The material presented by Fleming appears to be compatible with this interpretation. Unquestionably, there are many documents which refer to a 'wish of the Führer', showing that Himmler, Heydrich, and the others believed that they were acting in accordance with Hitler's intentions, and in fact, he did not interfere in Himmler's dirty business, though officially endorsing no more than the progressing deportation programme. There are compelling indications that no formal order by Hitler to implement the programme for the liquidation of European Jewry ever existed. The famous order presented by Göring to Heydrich on 31 July 1941 had been drafted by Heydrich himself and there was no connection with any initiative by Hitler.[20] In fact, this document shows that the *Reichssicherheitshauptamt* had finally succeeded in acquiring exclusive competency in the Jewish question. The aforementioned programme for a 'definitive solution to the Jewish question' was formulated when the Nazi leadership expected to defeat the Soviet Union within a few weeks and in any event before the winter of 1941. Hence, the preparations for a definitive solution mentioned in Göring's order refer to a post-war situation and are not, therefore, identical with the actions against the Russian Jews performed at the same time by the *Einsatzgruppen*. Although the Holocaust, i.e. the systematic liquidation of the Jews living within the German sphere of influence, started with

[19] Fleming, *Hitler und die Endlösung*, 32; Broszat, 'Genesis der "Endlösung" ', 763; cf. William Carr, *Hitler, A Study in Personality and Politics* (London, 1978), 72, 76.

[20] Compare Hilberg, *Die Vernichtung der Juden*, 283.

the mass killings in the Soviet Union performed under the pretext of unavoidable military measures, in particular in order to prevent further partisan warfare.

This being so, the order of July 1941 has no connection with an alleged order by Hitler which many experts attribute to November 1941. Unquestionably, late in 1941 a fundamental change in the anti-Jewish strategies occurred which was reflected in the fact that the security police abruptly stopped any further Jewish emigration from German-occupied territories. Up to the summer of 1940, the leading figures in the Third Reich, Heinrich Himmler, Reinhard Heydrich, Hermann Göring, Martin Bormann, and others, did not perceive any other solution to the Jewish question than the enforced emigration systematically pursued by Adolf Eichmann, who became the leading personality in the Reich Emigration Office, established as part of the *Reichssicherheitshauptamt* in Berlin.

4. The road to Auschwitz

Karl A. Schleunes has pointed out that the implementation of Nazi anti-Jewish policy until 1941 did not emerge from any clear-cut plan but followed a trial-and-error method.[21] Even after all responsibility had been concentrated in the hands of Heydrich, individual activities by the *Gauleiter*, such as the expulsion of Jews from the Palatinate and the Saar to France in 1940, were still connected with the policy of enforced emigration fostered by Heydrich and Eichmann. The *Gauleiter* rivalled each other in the ambition of getting rid of the Jews in their respective districts; these served as an important impetus for more radical procedures against German Jewry during the two years preceding the Wannsee Conference in January 1942. The war and the occupation of Poland, as a consequence of which the number of Jews under German rule now rose to more than three million, meant that the emigration scheme no longer offered a way of solving the 'Jewish question'. Hence it was replaced by the concept of a Jewish reservation to be established on the Eastern border of the *Generalgouvernement*. This 'solution', supported by the East Ministry, Eichmann, and many others, failed,

[21] See K. A. Schleunes, *The Twisted Road to Auschwitz. Nazi Policy Toward German Jews 1933–1939* (Chicago, 1970), 258.

partly because the technical preconditions for such a programme were completely lacking.

For this reason Eichmann fervently supported the Madagascar Plan, which was seriously pursued until February 1942, when the expectation that the British would be compelled to accept Hitler's peace terms was shattered. All this shows that the vague concept of achieving a territorial final solution was originally envisaged for the period after the war. The situation changed in two ways. On the one hand the killing operations carried out by the *Einsatzgruppen* offered an opportunity to get rid of those Jews who were not or no longer fit for work. On the other hand, the Madagascar Plan and similar colonial solutions became obsolete. Their existence had been the precondition for Heydrich's order in September 1939 to deport the Jewish population living in the annexed Polish territory to the *Generalgouvernement*, because this contradicted the comprehensive settlement projects cherished by Heinrich Himmler. At that time Heydrich made a clear-cut distinction between a preliminary *ad hoc* plan and a definitive solution which was not to be carried out in the near future.

It is of the utmost importance that this first systematic deportation measure was inaugurated in the context of creating space for the German minorities in the Soviet Union who, according to the provisions of the German—Soviet Non-Aggression Treaty, were to be repatriated and brought to the Warthegau and other parts of the annexed former Polish territories. The fate of the Jewish population was, therefore, directly connected with Himmler's massive settlement plans in Eastern Europe which climaxed in the later 'General Plan East', endorsed by Himmler in his capacity as Reich Commissioner for the Strengthening of the German Folkdom (*Reichskommissar zur Befestigung des deutschen Volkstums*). Together with the first unsystematic deportations to the Łódź ghetto, which became the transfer-centre for deported Jews from the Reich and the annexed territories, this programme, accompanied by the compulsory allocation of the Polish Jews to overcrowded ghettos with dreadful housing conditions, without adequate sanitation, and with extreme unemployment, created utterly intolerable living-conditions, particularly in Łódź.

The higher police and SS officer there complained in July 1941 in a letter to Eichmann about the catastrophic conditions in the Łódź ghetto and added the remark that serious consideration should be given to the idea that the most humane solution might be to kill the Jews, since they were no longer fit for work, and to do so by some rapidly killing chemical means. Martin Broszat, as well as Kurt Pätzold, have drawn attention to this detail.[22] The proposition that killing the Jews was a more 'humane' measure than letting them die of hunger and diseases is not isolated. The diabolical circle of fascist policies to create deliberately intolerable conditions and states of emergency and then to use them to legitimize even more radical steps becomes quite clear in this respect; but this mechanism does not mean that the outcome is preconceived from the start. The moral responsibility for the escalation of crime is evident, and it is implicit within the first steps which are deliberately taken. That they turn into a self-propelling process, thereby also producing secondary legitimations, although of an extremely dubious and non-rational kind, characterizes Nazi politics. In the *Generalgouvernement* deviated trains with deported Jews in 1940, who had no drinking-water or food, had already given rise to the idea that in the last resort it would be more humane to liquidate the victims immediately. This was also the general experience as regards conditions within the occupied parts of the Soviet Union. But there was one basic difference: the partial liquidation of Jews from the Reich and annexed territories represented a new step because in this context the familiar justification of the need to destroy Bolshevik resistance did not fit. In particular, the intolerable living-conditions which arose from unco-ordinated deportations gave rise to pseudo-justifications of the policy of systematic liquidation. Inhumanity had to be disguised as humanity before any programme could be implemented which could then be directed against all the deportees, including women and children, without distinction.

The change in the conception of the 'solution to the Jewish question' in the *Reichssicherheitshauptamt* did not occur abruptly. The genocide started with partial liquidations, and several times the deportation measures had to be interrupted. It was not until the spring of 1942 that the systematic policy of genocide became reality. Before that, in the late autumn of 1941, Action T4 had been put into operation; this had too the purpose of avoiding the popular unrest that had occurred in conjunction with the disorganized mass killings in Riga, whose victims were also German Jews deported from Berlin and other cities.

The Wannsee Conference, whose postponement Fleming explains by arguing that experts on annihilation techniques were not available in

[22] Broszat, 'Genesis der ''Endlösung'' ', 749 and S. K. Pätzold, 'Von der Vertreibung zum Genocid. Zu den Ursachen, Triebkräften und Bedingungen der antijüdischen Politik des faschistischen deutschen Imperialismus', in D. Eichholz and K. Gosweiler (eds.), *Faschismusforschung. Positionen, Probleme, Polemik* (Berlin (DDR), 1980), 197 ff.

December, is generally regarded as the decisive turning-point.[23] Hans Frank noted that partial liquidations had to be managed, and he added that this was supposed to be performed 'in conjunction with the great measures' to be discussed 'at Reich level'.[24] This implies that, at the moment, the technical facilities for the imminent implementation of genocide was still not available; it was symptomatic that one of Eichmann's first tasks in conjunction with the preparations for the Wannsee Conference consisted of checking the killing practices for the *Reichssicherheitshauptamt* which had been developed by members of the euthanasia programme, as well as the gassing techniques which were installed at Auschwitz-Birkenau in order to kill Soviet prisoners of war there.

At the Wannsee Conference the decision was made to include all European Jews in the deportation programme; it was still left undecided how half- and quarter-Jews, and Jews living in so-called privileged marriages, should be treated. The annihilation programme still appeared to be quite vague and Heydrich's remarks could be interpreted in different ways, although he mentioned the necessity of the later extermination of those deportees who might survive the process of annihilation through work. The fictitious inclusion of the Jews in a programme of compulsory labour—*Arbeitseinsatz*—constituted a psychological chain leading from the emigration solution through the reservation solution to genocide. It implied the unavoidable liquidation of that part of the Jewish population no longer capable of working under the unimaginably terrible conditions which the German authorities had already introduced in Eastern Europe, and which were officially designated as necessities of war.

It is not surprising that, given the conditions, the programme of using the Jews for compulsory labour was immediately dropped, owing to the lack of any co-ordinated preparation, in favour of systematic mass killings, even though the fiction of the *Arbeitseinsatz* still remained the official line. Even anti-Semitic extremists such as Frank and Kube protested when the automatic liquidation process included their urgently needed Jewish labour force, and the same was true when the Lublin activities of the Reich Economic Main Office of the SS were suddenly disrupted by the deportation of the Jewish workers to the annihilation camps.

[23] See Fleming, *Hitler und die Endlösung*, 105. It is not likely at all that at the Wannsee Conference annihilation techniques were discussed, as Fleming assumes.

[24] W. Präg and W. Jacobmeyer (eds.), *Das Diensttagebuch des deutschen Generalgouverneurs in Polen 1939–1945* (Stuttgart, 1975), 457 f.

Furthermore, it is important to realize that a connection also existed between the fate of the Soviet prisoners of war and that of the European Jews. Auschwitz-Birkenau had been developed into a huge armament centre relying on the labour potential of the prisoners. The fact that only a small number of the Soviet prisoners of war survived the brutal treatment of the German military authorities led even before the Wannsee Conference to the order by Himmler to deport the German Jews to the SS-owned concentration camps. The Birkenau camp, where annihilation techniques had already been developed before this decision, was then used for the implementation of the genocide programme, whilst territorial solutions were dropped, because they would have interfered with Himmler's gigantic settlement plans in the East. Official language still used the term *Arbeitseinsatz* to conceal the reality of murder.

To understand why compulsory labour served as a smoke-screen for extermination one must be aware that Germany and Europe had been accustomed to the existence of labour-camps of all kinds, ranging from voluntary labour to the concentration and extermination camps. The existence of the camps, the separation of families and sexes, the transport of men over the continent all existed even before the war and constituted something of a 'second civilization'. The extermination camps like Auschwitz, Maidanek, Sobibor, Treblinka, and many others were masked as labour-camps. More important, however, is the technique of first pressing the Jewish population of the occupied territories into camps of this kind, and afterwards deporting them 'to the East'. That partly explains why open resistance to the deportation was a rather exceptional phenomenon. Together with the ideology of 'national labour', which traces back to German nationalism of the nineteenth century and found its most cynical expression in the inscription at Auschwitz 'Arbeit macht frei',[25] this phenomenon was an important factor in the widespread moral indifference which made the implementation of the Holocaust possible.

It is not necessary to assume that the policy of genocide derived from a special order by Hitler. On the contrary, there are many arguments to show that this was not the case. Any order such as this, even if delivered orally, would have contradicted the prevailing fiction of systematic compulsory labour together with the calculated annihilation through labour. This did not suit the interests of the dictator, who would then

[25] See Frank Trommler, 'Die "Nationalisierung" der Arbeit', in R. Grün and J. Hermann (eds.), *Arbeit als Thema in der deutschen Literatur vom Mittelalter bis zur Gegenwart* (Königstein, 1979), 102–25.

have been confronted with the necessity of differentiating between *Arbeitseinsatz* and liquidation. Conversely, Hitler circumvented a distinction in this respect as the treatment of Jewish labour in the *Generalgouvernement* shows.

The absence of any formal order also explains why interests opposing the destruction of the Jewish labour-reservoir did not find anyone in charge. Even Frank did not dare to break the official taboo when he met the Führer. Hitler himself did not perceive the necessity of identifying himself directly with the genocide measures. There can be no doubt that, at any stage in the development of anti-Jewish actions, the dictator favoured radical solutions or that he would not hesitate to call for the killing of entire populations, as in the case of Leningrad. It is significant that he also took up the apologetic argument of using more 'humane methods', as in his aforementioned speech to the generals in 1944, but he still spoke in terms of a potential, not of an actual procedure.

It was Himmler and his subleaders who provided the means and the framework for the Holocaust. What appeared to Hitler a millenarian ideas was changed by Himmler, as his Posen speech reveals, into an actual and systematic policy.[26] Hitler was an outstanding example for the overall tendency, which he had pressed on the party in the early twenties, to repress those parts of reality which might question the reliability and coherence of the 'National Socialist idea', whose indispensable standard-bearer Hitler claimed to be and to remain. A similar attitude is to be found among nearly all the Nazi leaders who participated in a process of collective repression whilst being part of the machinery of the Holocaust. The general moral escapism characteristic of the members of the Nazi élite and of the great many people who, directly or indirectly, served Himmler's machinery, is certainly not an invention of the Nazi movement. Behavioural patterns like this can perhaps be traced back to the nineteenth century and they seem in particular to be typical of the German élites, as Rainer C. Baum has tried to show in order to explain the process of adaptation to moral indifference,[27] of which the Holocaust is only one, but the most terrible, example. The Holocaust, therefore, was a traumatic experience for the victims, for those who survived, but in a certain sense also for those who did not or were not able to protest against it. Obviously, the ideological background—racial

[26] See n. 6. Compare Broszat, 'Genesis der "Endlösung" ', 773 ff. and 'Soziale Motivation und Führerbindung', 408.

[27] See R. C. Baum, *The Holocaust and the German Elite. Genocide and National Suicide in Germany 1871–1945* (London, 1981), esp. 294 ff.

anti-Semitism, but also all sorts of ideologies excluding ethnic, social, or religious minorities from the right of equality and independent human existence—proved to be an important precondition for what happened. The role of bureaucratic structures of a secondary character—as for example the *Reichssicherheitshauptamt*—may have in different forms counterparts even in Western democratic societies. One lesson one can draw is to avoid the dissolution of the political process and the concealment of responsibility in an anonymous political structure. But apart from that the anthropological dimension is not to be forgotten: the danger inherent in present-day industrial society of a process of becoming accustomed to moral indifference in regard to actions not immediately related to one's own sphere of experience. The attitude of the German population, which in its majority did not face up to the consequences of racial anti-Semitism, reflects this observation: the almost complete segregation of the Jews supported moral indifference and collective repression. In this respect, recalling Helmuth James von Moltke's dictum that it was necessary to restore the image of man in the hearts of our fellow-citizens,[28] and Adam von Trott's endeavour to preserve the role of human personality in the advanced industrial society, remains an indispensable task of our times.

[28] Letter by Helmuth von Moltke to Lionel Curtis on 18 Apr. 1942, in M. Balfour and J. Frisby, *Helmuth von Moltke: A Leader Against Hitler* (London, 1972), 184.

8

BRITAIN AND THE THIRD REICH

PETER LUDLOW

THE discussion of British policy towards the Third Reich, at least amongst professional historians, has moved a long way beyond the grand simplicities of the 1940s and 1950s. With such a strongly defined and entrenched orthodoxy, a revisionist reaction was almost inevitable. The 'guilty men' could not possibly have been as black as they were painted: the heroes must have had their weaknesses. The debate that ensued has not, however, been simply or even mainly one between revisionists and anti-revisionists, though there were and there are examples aplenty of both. What has happened instead is that the issues have become much more complex, as each new contribution has thrown fresh light on a neglected actor or a forgotten consideration.[1]

British policy towards the Third Reich, it is now clear, cannot be seen apart from British policy towards the Soviet Union,[2] or the Commonwealth,[3] or the United States,[4] or France.[5] It would be rash too to ignore the economic constraints under which policy-makers laboured: the priorities and fragility of the recovery programme of the National Government, the weakness of Britain's external financial position which had been so dramatically displayed in the run on the pound in 1931, the shortages of skilled labour which would have quickly manifested themselves if rearmament had been speeded up.[6] Other historians have shown the importance of military doctrines or prejudices: of fears and delusions about the implications of air-power: of the determination to avoid at all

[1] For a very full bibliography, cf. G. Schmidt, *England in der Krise, 1930–37* (Opladen, 1981).

[2] G. Niedhart, *Großbritannien und die Sowjetunion, 1934–39* (Munich, 1972).

[3] R. Ovendale, *'Appeasement' and the English-speaking World* (Cardiff, 1975).

[4] e.g. D. Reynolds, *The Creation of the Anglo-American Alliance 1937–41* (London, 1981).

[5] e.g. R. A. C. Parker, 'Great Britain, France and the Ethiopian Crisis, 1935–36' *English Historical Review* (1974), 293 ff.; J. T. Emmerson, *The Rhineland Crisis* (London, 1977); *Les Relations franco-britanniques, 1935–39* (Paris, CNRS, 1975).

[6] e.g. B. J. Wendt, *Economic Appeasement* (Düsseldorf, 1971), G. C. Peden, *British Rearmament and the Treasury* (Edinburgh, 1979); R. A. C. Parker, 'British Rearmament 1936–39: Treasury, Trade Union and Skilled Labour', *English Historical Review*, (1981), 306 ff. B. M. Rowland (ed.) *Balance of Power or Hegemony: The Interwar Monetary System* (New York, 1976).

costs another war in the trenches: of the low opinion of French military capabilities.[7] According to another view, Hitler's programme was arguably less important in determining the pursuit of appeasement, and still more its abandonment in 1939, than the electoral interests of the Conservative Party.[8]

As the range of issues that have to be considered has widened, the number of actors who are seen to have played a role in the story has increased. The gallery of guilty men has been almost blotted from view, as each new book has brought into the forefront yet another individual or group. Besides the original cast of Chamberlain, Baldwin, Simon, and the Cliveden set, we now have the Treasury, the Chiefs of Staff, the appeasers in the Foreign Office, the Tory Right and the Christian Left, the City, Northern manufacturers with a strong interest in exporting to Germany, Dominion Prime Ministers and High Commissioners, the 'pacifists', and by no means least, public opinion, as it was or as it was perceived to be.

And so we might go on. As each fresh ingredient has been added, and each new stir given to the pot, the mixture has become thicker, less transparent, less digestible. The appeasers were neither as stupid nor as warped as they were held to be: the anti-appeasers had no monopoly of wisdom or foresight. Few if any British historians would be rash enough now to offer, as Keith Robbins did a few months before, and Keith Middlemass did only slightly longer after, the official archives were opened, a systematic account of 'Munich' or 'the diplomacy of illusion'.[9] Some foreign historians have been braver, but the results, though monumental, have been scarcely reassuring. Telford Taylor seems to have decided that the best way forward was to ignore the professional debate altogether,[10] while Gustav Schmidt, who appears to have read absolutely everything, both secondary and primary, pushes his readers remorselessly deeper and deeper into a wood from which there is no exit, and into which no light penetrates.[11]

Is it, therefore, quite impossible to arrive at some general judgements on British policy towards the Third Reich, which do not deny the

[7] Michael Howard, *The Continental Commitment* (London, 1972); B. Bond, *British Military Policy between the Two World Wars* (Oxford, 1980).

[8] M. Cowling, *The Impact of Hitler: British Politics and British Policies* (Cambridge, 1975).

[9] K. G. Robbins *Munich, 1938* (London, 1968); K. Middlemass, *The Diplomacy of Illusion* (London, 1972).

[10] T. Taylor, *Munich: The Price of Peace* (London, etc., 1979).

[11] See n. 1.

numerous nuances that study after study have introduced into the discussion, but which nevertheless distinguish between the important and the less important? Must we simply accept that in a story that is, whichever way one looks at it, a story of failure, failure to prevent a war and failure to win it except in a manner and at a cost that left both this country and the Continent poorer and weaker, the leading actors could not have done other than they did, so deep were the currents of history in which they were caught up? Faced with the daunting task of sorting out priorities and making judgements, it would perhaps be tempting to believe that neither is possible, to emulate Lord Franks and his colleagues by rehearsing the evidence with meticulous care, but eschewing any comments that might seem to apportion responsibility or blame to those involved.[12] It would be tempting, but timid, and furthermore unnecessarily timid. Complex the story undoubtedly is, but several recent studies do point towards certain conclusions both about the policies that were followed, and those that might have been considered, that strike at the heart of the matter.

Before moving on to a discussion of some of the principal policy errors towards Germany committed by the British leaders in the 1930s, however, it seems appropriate to look for a moment at one area of policy in which, it has frequently been alleged, the British authorities made serious errors of judgement which, had they been avoided, might have prevented the war altogether or at the very least hastened its end. I refer of course to the British government's handling of opposition elements in the Third Reich. The arguments advanced by these critics of British policy are relatively straightforward.[13] Had the British government accepted the advice of German opponents of the regime in 1938, and stood firm against Hitler during the Czech crisis, Hitler would have been overthrown and his regime replaced by one that would have co-operated with the British in reconstructing Europe along peaceful lines. Later, during the war itself, the behaviour demanded of the British, namely a more generous attitude towards the 'other Germany' both in public propaganda and secret replies to secret feelers, was almost the reverse, but the case is the same. Had the British acted as advised, Hitler's opponents would have removed him and made peace.

[12] *Falkland Islands Review. Report of a Committee of Privy Councillors* (Chairman: Lord Franks). Cmd. 8787 (London, IMSO 1983).

[13] The *locus classicus* remains H. Krausnick and H. Graml, 'Der deutsche Widerstand und die Alliierten', in *Vollmacht des Gewissens*, ii (Frankfurt and Berlin, 1965).

It would of course be highly appropriate in a course of lectures devoted to the memory of a notable opponent of the regime, if this case could be sustained. Unfortunately, however, it cannot. This is not, it need hardly be said, tantamount to saying that Maurice Bowra and his Oxford contemporaries were right to behave as pettily as they did towards Trott,[14] nor is it equivalent to claiming that the British government's policy towards the opposition in the Third Reich was above criticism. Neither case is true. But there is little or nothing in the evidence that we now have about opposition in the Third Reich to indicate that a different policy on the part of the British government in 1938 would have galvanized it into a force capable of removing Hitler, and there is no evidence at all in the British archives of which I am aware—rather the reverse—to suggest that the British authorities would have been justified in making the possibility of a coup in Germany the ground for resisting Hitler in the autumn of 1938.

This is not the moment to enter into a lengthy discussion of the historiography of the German resistance since the Second World War.[15] Suffice to say, that for reasons that were in most cases both honourable and legitimate, German and non-German advocates of Hitler's opponents, many of whom had paid for their opposition with their lives, tended to exaggerate both the earliness with which a distinct political opposition emerged, and its cohesiveness and effectiveness. The material that we have at our disposal, not least in the British archives themselves, suggests a much more complex evolution, in which those who were well before the end of their lives out-and out-opponents of the regime, and who throughout were repulsed by the latter's policies towards the Jews, the churches, and its other victims, were slow to abandon the feeling that the ship of state, even with the Nazi party in the ascendant, might still be steered on to a safer course, and still slower to come to terms with the ethical and organizational implications of resistance in a police state.

The case of Carl Goerdeler, who was, during the pre-war period at least, the most frequent informant that the British government had within opposition circles, is illustrative in this respect.[16] Goerdeler first appears in the British archives after 1933 as a servant of the regime: as

[14] C. Sykes, *Troubled Loyalty* (London, 1968) 262 ff.

[15] The most comprehensive survey is Peter Hoffmann, *The History of the German Resistance 1933–45* (London, 1977).

[16] Cf. G. Ritter, *Carl Goerdeler und die deutsche Widerstandsbewegung* (Stuttgart, 1956).

Reichskommissar for prices to be precise. As such, and still more as mayor of Leipzig, he was to be anything but a spineless collaborator, but even when he went into opposition and embarked in 1937 on the earliest of his foreign journeys, the picture that he gave of the politics of the Third Reich and of the position of people like himself in it was by no means unambiguous.

In June 1937, for example, Waley, a Treasury official of whom I shall have more to say later, noted that Goerdeler, during the course of a conversation with him, had described the situation in the following terms: ' "there must" he said "be a trial of strength between the parties around Hitler, and certain influences"—he meant the Nazi bosses and he mentioned Himmler by name—"must be removed from power . . . ." General Göring he ranked amongst the angels and he said that "Hitler himself would be alright [sic] if the evil influences could be removed from his entourage." '[17] It was of course a tale that had been told to, and believed in, London often enough before, but the significant point about it is that Goerdeler went on speaking like this even though he had given up his official positions, and embarked on opposition-like activities. Only a week or two after this particular conversation, for example, Goerdeler spoke to another English contact, A. E. Barker of *The Times*, about his efforts to push Fritsch and the army into a military coup.[18]

As the months passed, Goerdeler's criticisms of the regime became more and more outspoken, but his credibility, and the credibility of those whom he represented, did not grow correspondingly—for at least three reasons. In the first place, despite his increasingly radical rejection of the regime and its excesses, he could not prevent himself, perhaps unwittingly, from expressing doubts about the ability of any new government to eradicate Hitler's hold over popular imagination. As late as May 1939, for example, while promising that the opposition would 'liquidate the SS', he observed that 'it might be possible to retain Hitler if he would accept the position as a President in a gilded cage, but without control of policy and without command of the army.'[19] Secondly, Goerdeler, for reasons that are all too understandable, spoke too often and too optimistically about events that did not materialize. When he reported on his conversation with Fritsch about a coup in the middle of 1937, he admitted that Fritsch had argued that 'public opinion' was

[17] Public Record Office, FO 371/20732. C. 4714/165/18.
[18] Ibid C. 4882/165/18.
[19] FO 371/22973; C. 8004/15/18.

not yet prepared for such a development. Goerdeler nevertheless told his British contact that he felt that the dictatorship 'could not carry on for more than 6–8 months'.[20] Eight months later, however, it was not Hitler who fell but Fritsch. Sir Orme Sargent may have shown little sympathy for the awful dilemmas with which Goerdeler and his friends were confronted, when in 1939 he poured scorn on 'their intentions, their ability and their courage',[21] but it is none the less true that the optimism that Goerdeler almost had to maintain, if he was not to throw in the towel in desperation, made the claim that there were forces inside the Third Reich capable of overthrowing the Führer seem increasingly threadbare to those who listened or watched outside. His credibility was undermined still further by a third aspect of his lobbying on the opposition's behalf, namely the far-reaching character of the territorial and political demands that he implied the new regime itself would advance once Hitler was thrown out. This is a point that has been made with considerable candour by the German historian Hermann Graml, and it need not be laboured here.[22] In the light, however, of the nine-point programme that Goerdeler communicated to Ashton-Gwatkin in December 1938, and which contained, amongst some very constructive suggestions, demands that the British should back Germany in 'the liquidation of the Polish corridor' and that Germany should be allowed to concentrate her energies on 'developments on her Eastern frontier' including, albeit with the help of the British, the French, and the Americans, 'the restoration of a reasonable order in Russia'—a phrase which, as William Strang observed, 'smacked of *Mein Kampf*'—Strang's further comment was not entirely unreasonable: 'It is not easy to see what we should gain—except a year or two's uneasy quiet—by helping to put people like Goerdeler and the army' into Hitler's place.[23]

Goerdeler did not of course speak for the German opposition in any official sense: the opposition was not sufficiently well organized to *have* one spokesman. But in the light of this and similar evidence that the British received from senior critics of the regime, the scepticism of Strang and his colleagues is perhaps less surprising than the fact that, despite the considerable wariness of the professionals, the two leading figures in the first war cabinet, Chamberlain and Halifax, *did* for six

[20] FO 371/207232; C. 4882/165/18

[21] FO 371/22973; C. 8004/15/18.

[22] Cf. H. Graml, 'Resistance Thinking on Foreign Policy', in: H. Graml *et al.*, *The German Resistance to Hitler* (University of California, 1970).

[23] FO 371/21659; C. 15084/42/18.

months, between the outbreak of war and March 1940, do all that they could to encourage revolt in Germany. There is no time here to describe in detail the various contacts which Chamberlain and Halifax authorized through both official and unofficial channels within the first six months of the war.[24] They included the use of intermediaries like Dahlerus, Hohenlohe, Payne Best and Stephens at Venlo, the former Chancellor Wirth, Lonsdale Bryans, and, by no means last, Pope Pius XII. The cumulative effect of these efforts, however, was only to confirm the profound scepticism of the professionals and, belatedly, to undermine the credulity of Chamberlain and Halifax, who, despite the advice that they received from their officials and the French Government, continued, as Halifax himself put it, 'to hanker after a revolution in Germany' as the means of bringing an unwelcome war to a speedy conclusion.[25] The sequel in 1941–4, when anyway the international situation made unilateral action by the British Government extremely difficult if not impossible, is in these circumstances, however regrettable, not altogether surprising.

If the argument that the British government could have prevented the war by a more sensitive reaction to German opposition soundings in 1938 or 1939 is untenable, in the sense that it has been advocated by many apologists for the resistance movement since 1945, this does not mean that British policy towards dissidents in Germany was above reproach, or that different British policies towards the Third Reich in a more general sense might not have encouraged a different political evolution inside Germany. This latter point can only be understood, however, if it is seen in relation to a much profounder flaw in British foreign policy in the 1930s, namely the failure of policy-makers to accept, and even in certain cases to understand, the implications and constraints of the international system which industrialization and modern methods of warfare had created.

Shortly after Munich, while he was still basking in the praises that were being showered on him from all quarters, Neville Chamberlain wrote to his sisters with evident pleasure that he had just received an inscribed copy, from the author, of Professor Temperley's study of Canning.[26] In sending it, Temperley intended to underline the similarities between Chamberlain and his nineteenth-century hero. The worlds

[24] P. W. Ludlow, 'The Unwinding of Appeasement', in L. Kettenacker (ed.), *Das andere Deutschland im Zweiten Weltkrieg* (Stuttgart, 1977).

[25] Ibid, 40.

[26] University of Birmingham, Neville Chamberlain Papers (NC).

in which the two men operated were very different, but both had grasped the fundamental truth that the real art of the diplomat was to match his country's commitments and external relations to its resources. Temperley was of course principally impressed by the fact that Munich had prevented a war, which, like the Prime Minister himself, he did not feel that Britain was in any state to fight. Looking at British policy towards the Third Reich as a whole, however, and not just at the Munich settlement, which was its apogee, what is most striking is not that it was tailored to the modest amount of cloth that the British still had at their disposal, but that it cast Britain in a role in the international system that was far beyond its ability to perform.

Thanks to the work of several historians, Dr Peden most prominent amongst them, we now know that, contrary to the myths that their enemies propagated, Chamberlain and his Treasury team, who were dominant within the government throughout the period 1933–40, were amongst the first to identify Germany as the principal threat to British security, and to press for rearmament.[27] The subsequent implementation of the programme may, because of balance of payments problems and manpower shortages, have been slower than it should have been, or, with the imposition of war-like controls, could have been, but it is clear that without the prodding of Chamberlain and his men, it would have been even more dilatory.

Their laudable foresight in 1934 and subsequently did not prevent them, however, from committing the country whose liabilities they wanted to limit, but whose security they were willing to pay for, to a colossal gamble on the ability of this country alone to pacify or deter the dictator. Whether they were considering rearmament, discussing British economic policy in the knowledge that whatever war there might be would probably in the end be won by the country that had the soundest and most ample economic resources, or in the person of Chamberlain flying to Munich to sort out the Czech problem, Chamberlain and his advisers developed their ideas with only scant regard for the contributions that other countries might make in a common effort, and with the overriding intention of reducing their liabilities to those whom Hitler might devour. As time passed, Chamberlain's recklessness was undoubtedly fuelled by his conviction that in the last resort Hitler too was a rational man, who could be persuaded by reasonable concessions to adopt pacific policies. After reading Stephen H. Roberts's *The House*

[27] Cf. e.g. Peden, op. cit.

that Hitler Built at the beginning of 1938, for example, he noted in a letter to his sisters that if he accepted the author's conclusions, he would despair: 'But I don't and won't.'[28] His credulity, optimism, *naïveté*, call it what one will, was, however, more a consequence of the policy of appeasement than a cause of it. The origins of this policy and its fundamental error are to be found elsewhere, in the rejection of a multilateral approach to the management of interdependence, the acceptance of bilateral negotiations as the most promising and effective way of dealing with Hitler, or indeed with anybody else, and, above all, the readiness to defend national interests, at the expense of and without regard to the interests of other countries whose security and prosperity were closely entwined with those of Britain.

In October 1931, Orme Sargent wrote in a Foreign Office minute:

> The fundamental cause of the present 'crise de confiance'. . . is (the) duality in the international relationships of Europe . . . i.e. self-centred nationalism on the one hand, and international cooperation on the other. Whereas economics tend more and more towards internationalism, politics . . . have been tending . . . towards an intensive nationalism. As long as politics and economics are thus tugging in opposite directions, it is difficult to foresee how any real stability and permanence can be reached in Europe. Still worse, the universal slump . . . has added strength to the narrow policy of nationalism instead of strengthening the forces of economic internationalism.'[29]

The tragedy of the 1930s is that far from playing a leading role in combating nationalism, whether political or economic, the British government chose to seek its economic salvation and eventually its political salvation, firstly, by burning its bridges to the powers on which, in the final analysis, both its prosperity and security depended, and secondly, by adopting the heroic but futile role of the arbiter of Europe. It was not, of course, alone responsible for the breakdown of efforts to develop multilateral machinery to cope with the problems of interdependence: multilateral diplomacy is always a messy and complicated process, even when all the parties to it have strong reasons for wanting it to work, which was certainly not true in the 1920s and the 1930s. Witness the role of the Banque de France, to take only one example, in the international monetary system in the late 1920s.[30] But although the British were not unique in their behaviour, their attitude had a special significance, because the country's underlying economic strength, its

[28] N. C. 18/1/1037, letter of 30 Jan. 1938.
[29] FO 371/15196; C. 7563/172/62.
[30] Cf. S. Clarke, *Central Bank Cooperation, 1924–31* (New York, 1967).

place in the international and Western European economic systems, and its political stability cast it in a leadership role, albeit as part of, and not apart from, a multilateral European system.

Illustrations of the National Government's tendency to consider Britain's interests as something different from those of the country's Continental neighbours during the confrontation with the Third Reich are so numerous that the choice of two or three examples is bound to appear arbitary. It seems sensible, however, to concentrate on aspects of British external economic policy, partly because, as has been noted earlier, Chamberlain and the Treasury were the dominant forces inside the Government in the 1930s, but still more perhaps, because the rich materials that have been so persuasively presented by Michael Howard and Brian Bond on British military policy in this period, and in particular on the extraordinary evolution of plans for a British expeditionary force on the continent, need to be set alongside less well-known material on Britain's exernal economic relations, if the underlying motives of British policy are to be understood.[31] I want to look briefly, therefore, at the commercial and payments agreement with Germany of November 1934, and at aspects of the negotiations and discussions that preceded and followed the conclusion of the Tripartite Agreement in September 1936.

The events that preceded the commercial and payments agreement between Britain and Germany of November 1934 have already been described by Bernd-Jürgen Wendt, in his study of economic appeasement, which is still by far the most thorough and perceptive discussion of the subject.[32] As Wendt shows, the negotiations began in September 1934, after several months of extremely strained relations between the two countries. By rigorously cutting back on the amount of foreign exchange available to German importers of 'non-priority' goods, the Reichsbank under Schacht had inflicted serious damage on a number of British exporters, particularly those trading in woollens, cotton, coal, and copper. Between 1933 and 1934, for example, British exports in cotton yarns dropped by 3½ per cent overall, but by 36½ per cent to Germany. A similar pattern manifested itself in other industries that were heavily committed to the German market, and their representatives were not slow to show their anger. The government could hardly, therefore, avoid action, and following a bitter exchange of notes in August 1934, Leith-Ross, the Treasury's leading international troubleshooter, was dispatched to Berlin to negotiate directly.

[31] See n. 7.

[32] B. J. Wendt, op. cit., part 1; see esp. ch. 9.

The Leith-Ross mission took place at a critical moment in the evolution of the Reich and British policy towards it. The initial 'trial' period was over and the British government, with Chamberlain, Warren Fisher, and the Treasury to the fore, was in the process of conducting a defence review in which Germany was identified as the most likely threat to British security.[33] Leith-Ross himself prefaced his first report to the Cabinet, of 28 September 1934, with a brief political analysis, which was designed to show that British policy would have to be formulated henceforth on the assumption that this unpleasant regime was here to stay. The German people, he wrote, are 'still hypnotised by Hitler's personality . . . Just as in America, Roosevelt has captured the popular imagination, however much the details of his policy may be criticised, so vast masses in Germany have pinned their faith to Hitler and regard him as the only hope in their difficulties'. Furthermore, even if Hitler were to be dropped, the alternatives, 'Göring or some other nominee backed by the Reichswehr', would be no better. Germany, he averred, was like a ship that was being steered by 'an experienced and rather intoxicated helmsman' under the close supervision of an 'armed guard' who might take over if the helmsman got too drunk but who would not alter course in any radical manner.[34]

Faced with this disagreeable fact, Leith-Ross recommended a strategy that was to be characteristic of British policy for the next four years. It was underpinned by three propositions:

(a) The regime will not be hurt by a policy designed to injure Germany's economy;

(b) We ourselves should suffer if we pursued such a policy;

(c) We must therefore 'maintain our trade with Germany, so long as we can do so without increasing our credit commitments to her'.

Given the political realities in Germany in late 1934, these arguments were in themselves not unreasonable. What was wrong was not the analysis, nor even in general terms the policy prescription, but rather the particular instrument that was chosen to implement it, namely a bilateral agreement. The economic agreement of November 1934, which has quite properly been linked with the naval agreement six months later, as one of the two pillars of British policy towards the Reich, was strictly in terms of national interest not at all disastrous. Anglo-German trade never flourished in the 1930s but it certainly improved from the trough

[33] Peden, op. cit., ch. 4.
[34] PRO CAB 24/250, cf. 218(34).

of 1934. To some extent, therefore, it can be notched up as yet another example of the 'sound management' of a National Government which steered the British economy through the economic crisis of the 1930s more successfully than any other European government. But like its economic policy as a whole, the agreement needs to be seen not simply in terms of its impact on the British economy, but in terms of its implications for the recovery of the international, and more particularly the Western European economy, and its significance to those who ruled in Berlin. As far as the former point is concerned, it was, as Leith-Ross himself admitted, like any bilateral agreement, 'destructive of international trade', while as regards the second point, it provided the earliest imporant clue, well before the more obvious breakdown of the Stresa front, that Britain was ready to negotiate its way to individual comfort and security, even if it meant injuring its neighbours and accepting the Third Reich's rules of the game.[35]

The same underlying approach was to appear even more clearly in the events that preceded and followed the conclusion of the Tripartite Agreement between Britain, France, and the United States in September 1936. This is not the occasion to go into a detailed account of international monetary relations in the inter-war period, nor even into the diplomacy and politics of the Tripartite Agreement itself. The former has been dealt with, impressionistically but brilliantly, in a book of essays edited by Benjamin Rowland, while the latter has been described by Stephen Clarke in a brief, but very useful sequel to a much longer study of Central Bank co-operation in the 1920s.[36] My concern here is simply to highlight some of the things that were said about the Tripartite Agreement at the moment that it was signed, and the rather less impressive realities that underlay the rhetoric.

As a declaration of intent by the three leading Western economic powers about the management of their currencies, the Tripartite Agreement seemed to many at the time to herald the beginning of the end of the economic nationalism that had wrought such havoc in the world economy in the first page of the decade. The impression that it was indeed a turning-point was encouraged by at least some of the principals themselves. Witness, amongst other things, the invitation to the Belgian Prime Minister, Van Zeeland, to carry out a wide-ranging study of the

[35] Cf. Wendt, op. cit.

[36] B. M. Rowland (ed.), op. cit. (n. 6); S. Clarke, *Exchange Rate Stabilisation in the Mid 1930s: Negotiating the Tripartite Agreement*, Princeton Studies in International Finance, No. 41 (Princeton, NJ, 1977).

contemporary ills of the international economic system, with a view to recommending ways in which, through international co-operation of the kind that had been achieved in the Agreement, the crisis could be overcome.[37] Viewed in another perspective, it was also represented, particularly but by no means exclusively by the French, as a symbol of Western solidarity that might encourage moderate elements in Germany to lead the Reich out of its self-imposed isolation, or if that failed, provide a basis for common action by the major democracies and any other smaller countries that might rally to their side. Morgenthau, the United States Secretary of the Treasury, described it as 'the greatest move taken for peace in the world since the World War'—a judgement not uninfluenced by Morgenthau's well-advertised desire to be seen as its principal architect—but as the following quotation from the French Minister of Finance of the time, Vincent Auriol, suggests, it was not an entirely atypical one:

> A joint declaration over the names of President Roosevelt, Secretary Morgenthau, Premier Baldwin, Chancellor Chamberlain and Premier Blum would give the world the first real evidence since the war that the three great powers are genuinely cooperating . . . A monetary peace on this foundation will be the basis for working towards economic peace and finally achieving political world peace.[38]

So much for the rhetoric. The reality was rather different. Though mouthing slogans about international co-operation in public most of those involved, but particularly, it must be said, Chamberlain and his colleagues at the Treasury, viewed it much more pragmatically, not to say cynically. The French, Chamberlain observed in a private letter, were overdoing it. And as for Morgenthau, 'he wanted to get on the telephone to me, but I wasn't having any. I prefer to use a longer spoon in supping with the Americans.'[39] This attitude was of a piece with the line that Chamberlain and the Treasury had taken almost throughout the negotiations that preceded the agreement, and were to take in the events that followed it. When the idea was first mooted, in 1935, the British, for a variety of predominantly technical reasons, enjoyed the initiative in exchange-rate management. Taking advantage of the stabilization of the

[37] For the origins of the Van Zeeland Mission cf. *inter alia* T.160 F14735/01 f.

[38] Cited at Clarke, op. cit. 38. For Morgenthau's motives, cf. above all his diaries in the Roosevelt Library at Hyde Park. Also the articles by Joseph Alsop and Robert Kintner in the *Saturday Evening Post*, 1 and 8 Apr. 1939.

[39] For British attitudes, cf. esp. T 160/845, F. 13640/2 and 840 F. 13427/8. For Chamberlain see NC 18/1/978.

dollar in January 1934, and still more directly of the French monetary law of June 1928 under which the Bank of France bought and sold gold without limit at a fixed price, they did very much what they wanted to through operations in francs, which because of the gold-standard arrangements holding the dollar and franc rates within narrow margins worked through to the sterling–dollar rate as well. Were Britain to go back to a fixed-rate regime—presumably against gold, like the other two powers—Chamberlain explained in the House of Commons in March 1935, the country would be exposed to policy changes in France and the United States that might force an increase in the Bank of England's discount-rate, restriction of enterprise, and increases in unemployment. Britain, he continued, was not 'in a position to take these risks' and he refused, as Chancellor, 'to put the pound at the mercy either of the franc . . . or of the dollar.'[40]

It was only the increasing inevitability of the franc devaluation and all that this might entail that persuaded Chamberlain and his officials to modify their rigid position even slightly, and it is all too clear from the internal memoranda that they wrote on the subject that there was no intention whatsoever of moving towards the kind of co-operation in monetary affairs that the Bank of England and the Federal Reserve had practised in the 1920s, and to which we became accustomed for three decades after the Second World War. Confirmation of the limited extent of their commitment can be found in their reactions and behaviour during two episodes that followed the Tripartite Agreement: the attempt by the Oslo States to encourage the liberalization of international trade through a concerted programme of tariff and quota reductions amongst themselves, and the discussion provoked by the Van Zeeland mission and, eventually, by his report.

The Oslo States, a group of small but wealthy European countries—Sweden, Norway, Denmark, Belgium, the Netherlands, Luxemburg, and Finland—had come together as a group for the first time in 1930 to make a common protest against the drift into protectionism that was already so obvious.[41] Their practical efforts to halt the breakdown of international trade in the first half of the 1930s were, however, limited, even though all of them, as export-orientated economies, suffered considerably, particularly at the hands of Britain, which was as it always had been the principal export market of each of them, but which was now

[40] Clarke, op. cit. 17.

[41] For a brief history of the Oslo States cf. Belgian Ministry of Foreign Affairs: Oslo Convention, 2470 *bis*, memorandum of 7 Sept. 1939.

much less open to their wares. In a decade in which imports in general fell as low as 18 per cent of British GNP, as compared with 31 per cent in 1913, and the Commonwealth share of these imports rose from a 1928 level of 23 per cent to 36 per cent in 1938, the Oslo States were amongst the principal victims of this politically induced structural change in the British economy.[42] For the first five years, however, they seemed to have accepted their fate with little more than routine protests. But with the conclusion of the Tripartite Agreement, several of the leaders of these states, and in particular the Dutch Prime Minister Colijn, seem to have decided that an opportunity now existed to mount a concerted campaign in favour of the liberalization of international trade.[43] A series of speeches, particularly by Colijn, in the winter of 1936–7 was followed in May by a meeting of representatives of the Oslo States in The Hague, where it was decided that the group would embark on a number of practical measures to encourage trade amongst themselves. The principal object of the exercise, however was, as Colijn himself explained in the opening speech at the conference, to encourage or even 'shame' some of the larger powers, and in particular Britain and the United States, into acting on the promises set out in the Tripartite Agreement. The initiative was none the less a total failure. Not only did the British fail to do what the Oslo States wanted: they also made it quite clear that they did not approve of efforts by the smaller states to dabble in matters that were the preserve of the large ones and, still more, to overcome their disadvantages of size by acts of practical solidarity. What they were proposing to do, an official memorandum of 1937 noted, was a threat to British rights under most-favoured nation agreements and to the Ottawa agreements, particularly as far as the latter concerned British agriculture.[44] Not surprisingly, perhaps in the face of this negative reaction from London, the Oslo initiative collapsed early in 1938.

The British government was equally lukewarm about another effort to find a way out of the international crisis, namely the Van Zeeland mission of 1937–8, even though in this case they had been party to the request to him to carry out the work. Details of Van Zeeland's efforts in 1937 and early 1938 need not detain us here.[45] Suffice to say that, assisted by a senior Belgian banker, Frère, he carried out a lengthy

[42] Schmidt 225.
[43] The fullest documentation is in Swedish: MFA, HP 64D, vols. xvi and xvii.
[44] FO 371/21081, N1354/230/63.
[45] On Van Zeeland in general, cf. T. 160, F. 14735/01 and F. 14735/03/1–4.

investigation into what most if not all European governments and the American administration believed to be the roots of the present crisis, and what measures they advocated to rescue the world from its predicament. The results of his enquiries were set out in a lengthy report at the beginning of 1938, together with a fairly significant number of practical recommendations, including some, like the proposal for the creation of something like the IMF, which were to be revived in the Second World War. As the British government was to some extent a sponsor of the initiative, it could scarcely be as discourteous towards its author as it had been for example towards Colijn and the other Oslo State prime ministers over their initiative. It is nevertheless quite clear from the documents that the most important figures in Whitehall, both in Cabinet and in the departments, did not like the idea of an initiative of this kind, and liked its specific proposals even less. Sir Frederick Phillips, for example, a key figure in British external economic policy, found it difficult to accept that 'the small countries', of whom he saw Van Zeeland as a representative, had anything remotely useful to contribute to international economic affairs.

> The smaller countries [he wrote in June 1937], should by all means be excluded until a settlement of the current international monetary and economic problems has been reached. They are merely a nuisance in monetary negotiations, and they are always quite content to do what they are told if only the main powers agree.[46]

Reactions to the substance of the report, when it eventually appeared, were also negative and, in early drafts at least, insulting. One by one the departments concerned, the Board of Trade, Agriculture, the Treasury, fired off broadsides against any measures that might endanger the Ottawa agreements, or British agriculture, or the management of sterling. Van Zeeland's call for currency stabilization, for example, was dismissed as a 'dangerous proposal'. 'Values attained by the free-play of the market . . . command more confidence than values depending on legislative enactment or agreement.' As for an IMF, it would almost certainly encourage countries already burdened with excessive foreign debt to incur further external obligations and delay the necessary shifts in exchange rates.[47]

These two phases in the development of Britain's external economic policy during the 1930s are only cited here as illustrations of a fundamental tendency in British foreign policy, which found its logical

[46] T. 160/84, F. 13427/8, memo of 7 June 1937.
[47] Departmental views are documented in T160, F. 14735/03/1 ff.

conclusion in Neville Chamberlain's policy towards Germany in 1937 and 1938. The National Government, or at least those who were the dominant forces within it, building on, but extending aspirations and prejudices that had been apparent already in the 1920s, reacted to the world crisis by turning away from international co-operation and foreign entanglements. The British recovery, which was significantly successful by international standards, was to be based on cheap money and protectionism, and it mattered little how either policy affected other governments, particularly Britain's neighbours in Western Europe.

As Hitler emerged, the strong men in the government, notably Chamberlain and the Treasury, were amongst the first to see and argue that the Third Reich constituted the principal threat to British security and that the country should rearm accordingly. The rearmament programme that they backed was, however, built very largely on the assumption that, even if Britain eventually had allies, the British contribution would be made from Britain, or by British ships patrolling the waters. Chamberlain justified his rejection of a Continental commitment by the argument that air power would be the determining influence in the next war and that an expeditionary force would, therefore, be unnecessary.[48] He was, however, as several episodes in the first winter of the war, notably in Scandinavia, were to show, more a politician than a strategist, and it would therefore be misleading to interpret the strategy that he and his colleagues at the Treasury endorsed and in the end imposed on the government solely or even mainly in terms of military calculations. It was, at another level, part and parcel of a systematic effort to play down Britain's obligations to 'foreigners', of the kind that we have already seen in the field of external economic policy. The same mentality was apparent in the British conversations with the Belgians and French after Hitler's invasion of the Rhineland in 1936. 'In the end', Chamberlain noted, 'we succeeded in bringing the French to reason', by which, as he went on to explain, he meant that the government had obtained the agreement of the other two governments that there would be no sanctions, military or otherwise, and that 'our commitments under Locarno had if anything been limited rather than extended'.[49]

The most striking illustration of this fundamental approach was, however, it need hardly be said, the attempt to buy Hitler off in 1937 and 1938. Chamberlain may by this stage have allowed his illusions about Hitler's rationality to win the upper hand, and he also had very serious,

[48] Cf. Peden, op. cit. 118–50, and Bond, op. cit.
[49] Chamberlain, NC 18/1/952, 21 Mar. 1936.

and justifiable, anxieties about British and French military capabilities, but the most striking characteristic of the 'new direction' which in his own eyes he introduced into the conduct of British foreign policy when he became Prime Minister in 1937, was its ruthless disregard of the interests of any other of the states concerned, if as was fairly obvious, their interests conflicted with his overriding desire to settle with Hitler once and for all. 'Of course they want to dominate Eastern Europe', he wrote after Halifax's return from his German mission in November 1937, and as long as they promised not to break the peace in securing their objectives, he for his part was not going to prevent them.[50] Germany, he assumed, or at least he hoped, would then settle down to the consolidation of its own imperial system, just as the British and to a certain extent the French were doing with theirs.

An adequate explanation of what one might describe as the 'neo-nationalism' of the dominant figures in the National Government in the 1930s would require a great deal more time than is still left in this lecture. In one perspective, it may be seen as a compound of numerous specific dislikes of individual countries or groups of countries: dislike of the French for the way in which they behaved during and after the First World War; distrust of the Americans, both as a threat to British leadership in the Anglo-Saxon world, and particularly in the Roosevelt era as a persistent menace to the successful development of the British government's economic recovery programme; contempt for the small states, or for the League of Nations, in which, it was frequently asserted, they enjoyed far too much influence; hostility towards Russia, for all that had happened since 1917 and might, if the Soviet rulers had their way, happen still nearer home in the future. Foreigners, it must be clear to anybody reading Chamberlain's private papers, were not simply different, but essentially untrustworthy and ignorant.

Tempting though it would be, however, to explore the theme of xenophobia in Chamberlain and many of his colleagues and contemporaries, it would be misleading to exaggerate the importance of individual preferences and prejudices. British nationalism in the 1930s, which coloured every aspect of the foreign and domestic policies of the government, must in the final analysis be seen in relation to a much profounder and more widespread uncertainty about the country's place in the international system. This uncertainly has rarely been more elegantly analysed than in a book by André Siegfried, the French

[50] NC 18/1/1030, 26 Nov. 1937.

political scientist, entitled 'England's Crisis' which was published in 1931, first in French, and subsequently in an extremely readable English translation. 'England', he wrote, 'has in a sense fallen between two stools, the European continent to which she does not belong, and the non-European world for which she has neither the youth nor the temperament. She is beginning to realise slowly and rather regretfully that her splendid isolation has come to an end . . . [and that] she will eventually have to enter into some international economic alliance.' Inevitable though the choice was, however, Siegfried did not believe that the British government of the day would hurry into making it. 'What is much more likely is that England will not choose at all. Faithful to her tradition and her genius she will hover between the two groups, without giving herself completely either to one or to the other. A European England is a dream, and a closed Empire a Utopia.'[51]

Siegfried's prediction, written just before the National Government took office, is in many respects an uncannily accurate description of the clumsy, arrogant, and at the same time fearful attempt by Chamberlain and his colleagues to evade the fundamental implications of the country's diminished power and its geographical position in an age of advanced technological warfare. To have made the choice—and there was in the end, given the proximity of Hitler's Germany, only one choice possible—would have implied a change in the structure of Britain's external relations, and by no means least, in the management of her own affairs, so far-reaching that it could scarcely be contemplated.

Just how radical the implications of a decision to enter into what Siegfried called an 'international alliance' were can be seen in the events of 1939–40, when, following the collapse of the policy of appeasement, Chamberlain's government was obliged to accept that Britain's security was in the last resort inextricably bound up with that of her Continental neighbours, particularly France, and that the only way in which Britain could be defended was in alliance with France, and if at all possible with others of the despised smaller states as well. As I have already described at some length elsewhere the 'unwinding of appeasement', I trust I may be forgiven for not providing a detailed description of how, eventually, at the beginning of March 1940, Chamberlain himself endorsed a Foreign Office minute calling for urgent consideration to be given to the possibility of creating an Anglo-French union after the war, as the nucleus of a new Europe, with which eventually, the smaller neutrals

[51] A. Siegfried, *England's Crisis*, tr. H. H. and D. Hemming, (London, 1931), 231, 250–1.

and even a democratic Germany might also be associated.[52] What I want to emphasize on this occasion, still more than I did in that essay, is the far-reaching and, from a bureaucrat's point of view, extremely uncomfortable character of the commitment that the British entered into with their change of policy. What was at stake, in the negotiations that went on throughout the twelve months from the spring of 1939 to the spring of 1940, was in the last resort the 'independence' which the National Government had fought so hard to maintain, at virtually any cost.

As so often, it was at the Treasury that the battle was fought most keenly and the issues at stake were as a result most clearly displayed. When talks between the British and the French about economic co-operation began, the Treasury maintained that 'existing channels' were already more than enough to conduct the low-key exercise in policy co-ordination which was all that they believed or wanted to believe would be necessary.

> In a perfect world [Waley wrote in 1939], no doubt our joint resources ought to be regarded as a common pool, and a system of priorities applying to both of our needs, ought to be drawn up. But in our view, the world is not likely to be sufficiently perfect for this, even in time of war, and if such a system were contemplated, the only result would be that the French would want us to sacrifice our own requests for theirs with the result of constant friction.[53]

This initial attempt to keep the French at bay was, however, doomed to failure and as the months passed, the British came under increasing pressure, not only from the French, but also from more enlightened sections of public opinion at home, to enter into a number of far-reaching economic agreements, including in the case of the Treasury a financial agreement that was signed in December 1939. The details of these negotiations, which involved, from the British Treasury's point of view, the excruciating sight of French ministers and officials demanding a more disciplined approach to prices and incomes in the UK, cannot, alas, be given here.[54] What they illustrate perfectly, however, is the fundamental challenge to national sovereignty represented by membership of an alliance. As the Cabinet noted after the financial agreement had been virtually tied up, 'the system of a common purse, if carried to its logical conclusion, would mean British interference in the details of

[52] Ludlow, 'Unwinding of Appeasement', 40.

[53] FO 371/22916, C. 7593/130/17.

[54] Some references in Ludlow, op. cit. 25 ff. For a much fuller documentation cf. the forthcoming volume, P. W. Ludlow (ed.), *The Wartime Alliance and the Development of European Integration and Cooperation* (European University Institute, Florence).

the French budget and French interference in the details of the British budget'.[55] Faced with this awful prospect, which, in the light of events in the last two years of the First World War, at least some Treasury officials with longer memories and experience must have been able to imagine all along, it is not surprising that many of those involved, ministers and officials, continued, to the very end of the alliance, to kick and scream against its constraints. It was the absolute antithesis of all that Chamberlain and his colleagues had stood for since 1931 both at home and abroad.

To sum up. What I have argued is that British policy toward the Third Reich—and by that I mean the policy of the principal figures in the government, notably Chamberlain and the Treasury—can only be understood if it is seen as the logical outcome of the neo-nationalism of the government that took power in 1931, and that this in turn must be interpreted as a last-ditch attempt to avoid the threat to national autonomy which full-hearted acceptance of Britain's reduced role and membership of an interdependent international system was bound to entail. British governments in the nineteenth century were in the fortunate position of being able to be both internationalist and nationalist, because essentially it was they who determined the rules of the international game. By the 1920s, the necessity for enlightened internationalism was all the greater, but its application in practice could only mean a diminution of national sovereignty. The domestic and foreign policies of the National Government of the 1930s were a forlorn attempt to escape from the realities of interdependence.

I have concentrated, as was only right and proper, on Britain in isolation. It is important and fitting in the present context to realize that the painful and ultimately futile experiment with nationalism that the British made in the 1930s was made, albeit in their own way, by every other major European government and, with even less hope of success, by most of the smaller governments as well. Britain's case was a special one, partly because it was still in international trade as in so many other respects a natural leader of the international community, and partly because it had many more commitments, both sentimental and material, outside Europe. But it was by no means unique. Hitler flourished, both domestically and internationally, because for a substantial part of the pre-war period many of those who could have opposed him, and who

[55] PRO, CAB 65/2, 7 Dec. 1939.

certainly disapproved of much that he did, were prevented from attempting radical measures against him because their understanding of the threat that he represented was warped by disconcerting similarities between his demands and their priorities and prejudices.

There were of course many, both inside and outside Germany, who saw the issues clearly and who spoke up for different policies. But if we look not at those who might have made policy, but at those who did, Chamberlain, Chautemps, and Daladier, or at those, in the German context, who had a reasonable prospect of taking over, Goerdeler, Beck, and Witzleben, not to mention the leaders of the lesser states, Nygaardsvold, Stauning, Leopold III, the overwhelming impression is of isolated individuals or national groups, locked in a learning process, during which they were unable to communicate effectively with one another, and at the end of which, when the separate approaches to European problems that they had pursued had ended in disaster, neither they nor any of their successors in positions of responsibility had any longer the power to impose European solutions to Europe's problems.

SHORT BIBLIOGRAPHY

Adam von Trott

BOTHGE, EBERHARD, 'Adam von Trott und der deutsche Widerstand', *Vierteljahrshefte für Zeitgeschichte* 11 (1963).

DUFF, SHIELA GRANT, *The Parting of the Ways. A Personal Account of the Thirties* (London, 1982).

MALONE, HENRY OZELLE, 'Adam von Trott zu Solz: the Road to Conspiracy against Hitler' (Ph.D. thesis, University of Texas, Austin, 1980).

SYKES, CHRISTOPHER, *Troubled Loyalty: A Biography of Adam von Trott zu Solz* (London, 1968).

TROTT, ADAM VON, *Hegels Staatsphilosophie und das internationale Recht* (Göttingen, 1932).

—— ed. and intro.: *Heinrich von Kleist: Politische und journalistische Schriften* (Potsdam, 1935).

The German Resistance

BALFOUR, MICHEL, and FRISBY, JULIAN; *Helmuth von Moltke: A Leader Against Hitler* (London, 1972).

BONHOEFFER, DIETRICH, *Letters and Papers from Prison* (New York, 1972).

HOFFMANN, PETER, *The History of the German Resistance 1933–1945* (London, 1977).

LEBER, ANNEDORE, BRANDT, WILLY, and BRACHER, KARL-DIETRICH, *Conscience in Revolt: Sixty-four Stories of Resistance in Germany 1933–45* tr. R. O'Neill (London, 1957).

RITTER, GERHARD, *The German Resistance. Carl Goerdeler's Struggle Against Tyranny* (New York, 1958).

ROON, GER VAN, *German Resistance to Hitler: Count von Moltke and the Kreisau Circle*, tr. Peter Ludlow (London, 1971).

ROTHFELS, HANS, *The German Opposition to Hitler*, tr. L. Wilson (London, 1961, rev. 1970).

SCHEURIG, BODO, *Klaus Graf von Stauffenberg* (Berlin, 1964).

ZELLER, EBERHARD, *The Flame of Freedom. The German Struggle Against Hitler* (Coral Gables, Fla., 1969).

Hitler

BULLOCK, ALAN, *Hitler. A Study in Tyranny* (London, 1952).

FEST: JOACHIM, *Hitler: A Biography* (New York, 1970).

HAFFNER, SEBASTIAN, *The Meaning of Hitler* (London, 1979).

HITLER, ADOLF, *Mein Kampf*, tr. Ralph Mannheim, intro. D. C. Watt (London, 1969).

Hitler's Table Talk 1941–44. His Private Conversations, intro. H. R. Trevor-Roper (London, 1953).

JÄCKEL, EBERHARD, *Hitler's World View* (Cambridge, Mass., and London, 1981).

STERN, J. P. *Hitler. The Führer and the People* (London, 1975).

STONE, NORMAN, *Hitler* (London, 1980).

The Nazi State

BRACHER, KARL-DIETRICH, *The German Dictatorship: The Origins, Structure and Effects of National Socialism*, tr. Jean Steinberg (New York and Washington, DC, 1970).

BROSZAT, MARTIN, *German National Socialism 1919–1945* (Santa Barbara, 1966).

—— *Der Staat Hitlers: Grundlegung und Entwicklung seiner inneren Verfassung* (Munich, 1969).

BUCHHEIM, HANS, *Anatomy of the SS State* (London, 1968).

FEST, JOACHIM, *The Face of the Third Reich. Portraits of the Nazi Leadership* (London, 1970).

MOMMSEN, HANS, *Beamtentum im Dritten Reich* (Stuttgart, 1966).

O'NEILL, R. J. *The German Army and the Nazi Party* (London, 1964).

SCHOENBAUM, DAVID, *Hitler's Social Revolution. Class and Status in Nazi Germany 1933–1939* (New York, 1966).

SPEER, ALBERT, *Inside the Third Reich* (London, 1970).

WHEELER-BENNETT, JOHN W. *The Nemesis of Power; The German Army in Politics 1918–1945* (London and New York, 1954).

Nazi Foreign Policy

Documents on German Foreign Policy 1918–1945 Ser. C (London, 1957–); Ser. D (London, 1949–).

HILDEBRAND, KLAUS, *The Foreign Policy of the Third Reich* (Berkeley, 1973).

JACOBSEN, HANS-ADOLF, *Nationalsozialistische Außenpolitik 1933–1938* (Frankfurt am Main and Berlin, 1968).

RIBBENTROP, JOACHIM VON, *Zwischen London und Moskau. Errinnerungen und letzte Aufzeichnungen* (Leoni, 1953).

TAYLOR, A. J. P., *The Origins of the Second World War* (London, 1961).

WEINBERG, GERHARD L., *The Foreign Policy of Hitler's Germany. Diplomatic Revolution in Europe 1933–6* (Chicago,1970).

WEIZSÄCKER, ERNST VON, *The Memoirs of Ernst von Weizsäcker* (London, 1951).

WISKEMANN, ELIZABETH, *Europe of the Dictators 1919–1945* (London, 1966).

The Jewish Question and the Holocaust

ANDIES, HELMUTH, *Der Ewige Jude: Ursachen und Geschichte des Antisemitismus* (Vienna, 1965).

ARENDT, HANNAH, *Eichmann in Jerusalem. A Report on the Banality of Evil* (London, 1963).

COHN, NORMAN, *Warrant for Genocide* (London, 1966).

DAWIDOWICZ, LUCY, *The War Against the Jews 1933–45* (London, 1977).

HILBERG, RAOUL, *The Destruction of the European Jews* (London, 1961).

HÖSS, RUDOLF, *Commandant in Auschwitz* (London, 1959).

PULZER, PETER, *The Rise of Political Anti-Semitism in Germany and Austria 1867–1918* (London and New York, 1964).

REITLINGER, G., *The Final Solution* (London, 1968).

British Policy Towards the Third Reich

CARR, E. H., *The Twenty Years' Crisis 1919–1939* (London, 1939).

COWLING, MAURICE, *The Impact of Hitler: British Politics and British Policy 1933–1940* (Cambridge, 1975).

Documents on British Foreign Policy 1919–1939 (London, 1949–).

GILBERT, MARTIN, *The Roots of Appeasement* (London, 1966).

HOWARD, MICHAEL, *The Continental Commitment* (London, 1972).

MIDDLEMASS, K., *The Diplomacy of Illusion* (London, 1972).

MOMMSEN, WOLFGANG and KETTENACKER, LOTHAR, (ed.): *The Fascist Challenge and the Policy of Appeasement* (London, 1983).

OVENDALE, R., *'Appeasement' and the English-Speaking World* (Cardiff, 1975).

WENDT, BERND-JÜRGEN, *Economic Appeasement* (Düsseldorf, 1971).

WHEELER-BENNETT, JOHN W., *Munich. Prologue to Tragedy* (London, 1948).

The Third Reich in German and European History

ARENDT, HANNAH, *The Origins of Totalitarianism* (New York, 1951).

AYÇOBERRY, PIERRE, *The Nazi Question. An Essay on the Interpretations of National Socialism 1922–1975* (London, 1979).

BUTLER, ROHAN d'O. *The Roots of National Socialism 1783–1933* (London, 1941).

CRAIG, GORDON, *Germany 1865–1945* (Oxford, 1978).

DAHRENDORF, RALF, *Society and Democracy in Germany* (New York, 1967).

DEHIO, LUDWIG, *Germany and World Politics in the Twentieth Century* (New York, 1959).

HOLBORN, HAJO, *Germany and Europe* (New York, 1970).

MEINECKE, FRIEDRICH, *The German Catastrophe* (Cambridge, Mass., 1950).

RÖHL, J. C. G. *From Bismarck to Hitler. The Problem of Continuity in German History* (London, 1970).

STEINER, GEORGE, *In Bluebeard's Castle* (London, 1971).

TALMON, J. L., *The Origins of Totalitarian Democracy* (London, 1961).

TAYLOR, A. J. P., *The Course of German History. A Survey of the Development of Germany since 1815* (London, 1945).

INDEX